Disclaimer

The publisher of this book is by no way associated with the National Institute of Standards and Technology (NIST). The NIST did not publish this book. It was published by 50 page publications under the public domain license.

50 Page Publications.

Book Title: Information Technology: American National Standard for Information Systems: Data Format for the Interchange of Fingerprint, Facial, & Scar Mark & Tattoo (SMT) Information

Book Author: R. McCabe;

Book Abstract: Defines the content, format, and units of measurement for the exchange of fingerprint, palmprint, facial/mugshot, and scar, mark, & tattoo (SMT) image information that may be used in the identification process of a subject. The information consists of a variety of mandatory and optional items, including scanning parameters, related descriptive and record data, digitized fingerprint information and compressed or uncompressed images. This information is intended for interchange among criminal justice administrations or organizations that rely on automated fingerprint (AFIS) and palmprint identification systems or use facial or SMT data for identification purposes.

Citation: NIST SP - 500-245

Keyword: AFIS;fingerprint;grayscale;identification;image;mugshot;palm print;resolution;scanning

NIST Special Publication 500-245

ANSI/NIST-ITL 1-2000
Revision of
ANSI/NIST-CSL 1-1993 &
ANSI/NIST-ITL 1a-1997

Information Technology:
American National Standard for Information Systems— Data Format for the Interchange of Fingerprint, Facial, & Scar Mark & Tattoo (SMT) Information

National Institute of Standards and Technology
Technology Administration, U.S. Department of Commerce

NIST Special Publication 500-245

ANSI/NIST-ITL 1-2000
Revision of ANSI/NIST-CSL 1-1993 & ANSI/NIST-ITL 1a-1997

Information Technology:

American National Standard for Information Systems— Data Format for the Interchange of Fingerprint, Facial, & Scar Mark & Tattoo (SMT) Information

Sponsored by
Information Technology Laboratory
National Institute of Standards and Technology
Gaithersburg, MD 20899-8940

September 2000

Approved July 27, 2000
American National Standards Institute, Inc.

U.S. Department of Commerce
Norman Y. Mineta, Secretary

Technology Administration
Dr. Cheryl L. Shavers, Under Secretary of Commerce for Technology

National Institute of Standards and Technology
Raymond G. Kammer, Director

Reports on Information Technology

The Information Technology Laboratory (ITL) at the National Institute of Standards and Technology (NIST) stimulates U.S. economic growth and industrial competitiveness through technical leadership and collaborative research in critical infrastructure technology, including tests, test methods, reference data, and forward-looking standards, to advance the development and productive use of information technology. To overcome barriers to usability, scalability, interoperability, and security in information systems and networks, ITL programs focus on a broad range of networking, security, and advanced information technologies, as well as the mathematical, statistical, and computational sciences. This Special Publication 500-series reports on ITL's research in tests and test methods for information technology, and its collaborative activities with industry, government, and academic organizations.

This document is a contribution of the National Institute of Standards and Technology and is not subject to copyright. Any organization interested in reproducing ANSI/NIST-ITL 1-2000, *Data Format for the Interchange of Fingerprint, Facial & SMT Information*, is free to do so. However, there shall be no alteration to any of the material contained in the document.

Certain commercial entities, equipment, or materials may be identified in this document in order to describe an experimental procedure or concept adequately. Such identification is not intended to imply recommendation or endorsement by the National Institute of Standards and Technology, nor is it intended to imply that the entities, materials, or equipment are necessarily the best available for the purpose.

National Institute of Standards and Technology
Special Publication 500-245
Natl. Inst. Stand. Technol.
Spec. Publ. 500-245
78 pages (September 2000)
CODEN: NSPUE2

U.S. GOVERNMENT PRINTING OFFICE-
WASHINGTON: 2000

For sale by the Superintendent of Documents
U.S. Government Printing Office
Washington, DC 20402-9325

CONTENTS

Foreword vii

0 Introduction 1

1 Scope, purpose, and application 1

    1.1 Scope 1
    1.2 Purpose 1
    1.3 Application 2

2 Normative references 2

3 Definitions 3

4 Transmitted data conventions 3

    4.1 Byte and bit ordering 3
    4.2 Grayscale data 3
    4.3 Binary data 4
    4.4 Color data 4
    4.5 Scan sequence 4

5 Image resolution requirements 5

    5.1 Scanner resolution requirement 5
    5.2 Transmitting resolution requirement 5

6 File description 6

    6.1 File format 7
    6.2 File contents 7
    6.3 Implementation domains 8
    6.4 Image designation character (IDC) 8

7 Record description 9

    7.1 Logical record types 9
    7.2 Record format 11

8 Type-1 transaction information record 14

    8.1 Fields for Type-1 transaction information record 14
    8.2 End of transaction information record Type-1 16

9 Type-2 user-defined descriptive text record 17

    9.1 Fields for Type-2 logical records 17
    9.2 End of Type-2 user-defined descriptive text record 17

10 Type-3 low-resolution grayscale fingerprint image record 17

    10.1 Fields for Type-3 logical record 17
    10.2 End of Type-3 low-resolution grayscale fingerprint image record 19
    10.3 Additional low-resolution grayscale fingerprint image records 19

11 Type-4 high-resolution grayscale fingerprint image record 19

    11.1 Fields for Type-4 logical record 19
    11.2 End of Type-4 high-resolution grayscale fingerprint image record 20
    11.3 Additional high-resolution grayscale fingerprint images 20

12 Type-5 low-resolution binary fingerprint image record 20

12.1 Fields for Type-5 logical record 21
12.2 End of Type-5 low-resolution binary fingerprint image record 22
12.3 Additional low-resolution binary fingerprint image records 22

13 Type-6 high-resolution binary fingerprint image record 22

13.1 Fields for Type-6 logical record 22
13.2 End of Type-6 high-resolution binary fingerprint image record 23
13.3 Additional high-resolution binary fingerprint image records 23

14 Type-7 user-defined image record 23

14.1 Fields for Type-7 logical record 23
14.2 End of Type-7 user-defined image record 24
14.3 Additional user-defined image records 24

15 Type-8 signature image record 24

15.1 Fields for Type-8 logical record 24
15.2 End of Type-8 signature image record 25
15.3 Additional signature 25

16 Type-9 minutiae data record 25

16.1 Minutiae and other information descriptors 25
16.2 Fields for Type-9 logical record 27
16.3 End of Type-9 logical record 30
16.4 Additional minutiae records 30

17 Type-10 facial & SMT image record 30

17.1 Fields for Type-10 logical record 30
17.2 End of Type-10 logical record 36
17.3 Additional facial & SMT image records 36

18 Type-11 record reserved for future use 36

19 Type-12 record reserved for future use 36

20 Type-13 variable-resolution latent image record 36

20.1 Fields for the Type-13 logical record 37
20.2 End of Type-13 variable-resolution latent image record 39
20.3 Additional variable-resolution latent image records 39

21 Type-14 variable-resolution tenprint image record 39

21.1 Fields for the Type-14 logical record 39
21.2 End of Type-14 variable-resolution tenprint image record 42
21.3 Additional variable-resolution tenprint image records 42

22 Type-15 variable-resolution palmprint image record 42

22.1 Fields for the Type-15 logical record 43
22.2 End of Type-15 variable-resolution palmprint image record 45
22.3 Additional Type-15 variable-resolution palmprint image records 46

23 Type-16 user-defined testing image record 46

23.1 Fields for the Type-16 logical record 46
23.2 End of Type-16 user-defined testing image record 47
23.3 Additional Type-16 user-defined testing image records 48

24 Another individual 48

ANNEXES

Annex A - 7-bit ANSI code for information interchange 49

Annex B - Use of information separator characters 50

Annex C - Base-64 encoding scheme 51

Annex D - JPEG file interchange format 53

Annex E - Scars, marks, tattoos, and other characteristics 59

Annex F - Example transaction 64

FIGURES

Figure 1 – Byte & bit ordering 4

Figure 2 - Minutiae coordinate system 27

TABLES

Table 1   – Logical record types 6

Table 2   – Number of logical records per transaction 7

Table 3   – Information separators 12

Table 4   – Directory of character sets 16

Table 5   – Finger impression type 18

Table 6   – Finger position code & maximum size 19

Table 7   – Minutia types 26

Table 8   – Pattern classification 28

Table 9   – Type-10 facial and SMT record layout 31

Table 10 – Subject pose 32

Table 11 – Photo descriptors 33

Table 12 – Tattoo classes 34

Table 13a – Human tattoo subclasses 35

Table 13b – Animal tattoo subclasses 35

Table 13c – Plant tattoo subclasses 35

Table 13d	– Flags tattoo subclasses	35
Table 13e	– Objects tattoo subclasses	35
Table 13f	– Abstract tattoo subclasses	36
Table 13g	– Symbols tattoo subclasses	36
Table 13h	– Other tattoo subclasses	36
Table 14	– Color codes	36
Table 15	– Type-13 variable-resolution latent record layout	37
Table 16	– Type-14 variable-resolution tenprint record layout	40
Table 17	– Type-15 variable-resolution palmprint record layout	44
Table 18	– Palm impression type	44
Table 19	– Palm codes, areas, & sizes	45
Table 20	– Type-16 User-defined testing record layout	46
Table C1	– Base-64 alphabet	51

Foreword   (This foreword is not part of American National Standard ANSI/NIST-ITL 1-2000)

Federal, state and local law enforcement and related criminal justice agencies have procured or are in the process of procuring equipment and systems intended to facilitate the determination of the personal identity of a subject from fingerprint, palm, facial (mugshot), or scar mark & tattoo (SMT) images.

Fingerprint and palmprint images are acquired from flatbed scanners, Automated Fingerprint Identification Systems (AFIS), live-scan fingerprint and palmprint readers, and/or Image Storage and Retrieval (ISR) systems. An AFIS scans and stores the digital representations of fingerprint and palmprint images from inked or chemical cards. Live-scan readers scan the fingerprint and palmprint image data directly from the subject's fingers and hands. The scanned images are then processed to extract specific types of features from the images.

Sources used for the electronic capture of a subject's facial image (mugshot), and SMTs present on the subject's body include digital still and video cameras and other types of video recorders that capture images and produce digital image files directly from the subject's head and body. Scanners are used to digitize images from photographs, pictures, or sketches. The digital representations of these images consist of grayscale or color pixels depending on the application and equipment.

These digital images may be stored in a compressed or uncompressed form in an image storage and retrieval system (ISR) together with textual descriptive data and other information for each image. When required, specific images stored on a master file can be retrieved from the ISR and be incorporated as part of an electronic mugshot book, or an electronic line-up. Images selected may be the result of textual filters based on physical descriptive or information fields associated with each image. Stored SMT images can also be retrieved as part of an identification process. A human witness or an examiner using images retrieved from the system makes the confirming identification.

Features from the scanned fingerprint, palmprint, facial, or SMT images can be compared against a masterfile containing features extracted from previously scanned images. The result of these comparisons is a list of potential candidate identifications. A human examiner, using images retrieved from the system or fingerprint cards, then makes the final identification.

To effectively exchange identification data across jurisdictional lines or between dissimilar systems made by different manufacturers, a standard is needed to specify a common format for the data exchange. The data may either be scanned images or the processed minutiae extracted by the system.

The Information Technology Laboratory (ITL) of the National Institute of Standards and Technology (NIST) sponsored the development of this American National Standards Institute (ANSI) document using the Canvass Method to demonstrate evidence of consensus. This updated standard replaces ANSI/NIST-CSL 1-1993 and ANSI/NIST-ITL 1a-1997 that address the interchange of fingerprint, facial, and SMT data.

There are six annexes associated with this standard. Annex A is normative and contains the 7-bit ANSI code for information interchange. Annex B is informational and illustrates the use of the information separator characters. Annex C is normative and describes the base-64 encoding scheme. Annex D, a description of the Joint Photographic Experts Group (JPEG) File Interchange Format (JFIF), is normative and considered part of the standard. Annex E contains excerpts from the Eighth Edition (July 14, 1999) of the NCIC Code Manual for describing Scars, Marks, and Tattoos. Annex F is a comprehensive example of a transaction file containing fingerprint, mugshot, and palmprint logical records. The image exchange records contained in Annex F are formatted in accordance with this standard and are informative and not considered as part of the standard.

Suggestions for the improvement of this standard are welcome. They should be sent to the attention of R. M. McCabe, Fingerprint Standards, Information Access Division, Image Processing Group, NIST, 100 Bureau Dr, Mail Stop 8940, Gaithersburg, MD 20899-8940.

The following organizations recognized as having an interest in the standardization of the data format for the interchange of fingerprint, facial, and SMT information were contacted prior to the approval of this revision of the standard. Inclusion in this list does not necessarily imply that the organization concurred with the submittal of the proposed standard to ANSI.

AAMVANET, Inc.
AFR Consortium
AK Dept. of Public Safety
Abilene Police Dept. (TX)
Aware, Inc.
Barry Blain, Ph.D - Consultant
Biometric Identification, Inc.
Bundeskriminalamt (Germany)
Bureau ATF /Dept. of Treasury
CA Bureau of Criminal Information & Analysis
CA Dept. Motor Vehicles
CA DOJ; Bureau of Criminal Identification & Information
CA DOJ; Bureau Forensic Science
CT Dept. of Public Safety
CT Dept. of Social Services
Chicago Police Dept. (IL)
Cogent Systems
Comnetix Computer Systems, Inc.
Criminal Justice Coordination Council
Cross Match Technology, Inc.
Digital Biometrics, Inc.
Digital Descriptor Systems, Inc.
Drug Enforcement Administration
FBI / CJIS
FBI / LFPS
FL Dept. of Law Enforcement
Flagler County Sheriff's Office (FL)
Forensic Identification Services
Free Radical Enterprises
G. A. Thompson Co.
GA Bureau of Investigation
HI Criminal Justice Data Center
Heimann Biometric Systems GMBH
Higgins & Associates, Inc.
IA Division Criminal Investigation
IACP
IAI Latent Print Certification Board
IBM
ID Networks, Inc.
IL State Police – Information Services Bureau
IRS Forensic Science Laboratory

I/O Software, Inc.
Identicator Corp.
Identix
Imagis Technologies, Inc.
Indianapolis Police Dept. (IN)
Infineon Tech AG
KEU'PZ (Slovakia)
KS Bureau of Investigation
Kent Ridge Digital Labs.
King County Sheriff's Office (WA)
LA State Police
Lightning Powder Co., Inc.
Lockheed Martin Info. Systems
Los Angeles County Sheriff's Dept. (CA)
Los Angeles Police Dept. / Records & Identification (CA)
MD Dept. of Public Safety & Correctional Services
MD State Police
ME State Police
MI State Police
MN Bureau of Criminal Apprehension
MO State Highway Patrol
MTG Management Consultants
Miami-Dade Police (FL)
Miami Valley Regional Crime Lab. (OH)
Michael Fitzpatrick - Consultant
Mikos Ltd.
More Hits
NC State Bureau of Investigation
NEC Corporation
NEC Technologies, Inc.
NIST/OLES
NM Dept. Public Safety
NY State Division Of Criminal Justice
New York City Police Dept. (NY)
OK State Bureau of Investigation
Orange County Sheriff's Dept. (CA)
Orange County Sheriff's Office (FL)
PA Chiefs of Police Assoc.
PA State Police
PITO, UK
Performance Engineering Corp.
Pittsburgh Police Dept. (PA)

Portland Police Bureau (OR)
Printrak International
RCMP (Canada)
RoadRider Solutions Group, Inc.
SC Law Enforcement Division
STMicroelectronics, Inc.
Saber Imaging
Sagem Morpho, Inc.
Sagem SA
Salt Lake County Sheriff's Dept. (UT)
San Francisco Police Dept. (CA)
San Jose State University (CA)
TX Dept. of Public Safety

The Avator Group
Thompson, CSF
Toronto Police Service - Forensic ID
   (Canada)
US Dept. of Justice
US Postal Inspection Service
VA Judges' Chambers, Alex. Circuit Court
VA State Police
VT Forensic Lab. CJIS
Visionics Corp.
WA State Patrol
WI Dept. of Justice
WIN, Inc.

x

AMERICAN NATIONAL STANDARD   ANSI/NIST-ITL 1-2000

American National Standard
for Information Systems —

Data Format for the Interchange of Fingerprint, Facial, & Scar Mark & Tattoo (SMT) Information

0 Introduction

In 1993, ANSI approval was obtained for the "Data Format for the Interchange of Fingerprint Information" standard (ANSI/NIST-CSL 1-1993). The standard specifies formats to be used for exchanging fingerprint and other image data.

In 1997, ANSI approval was obtained for the expansion of that standard to include mechanisms for interchanging facial/mugshot image data and captured image data from scars, marks, and tattoos. That standard is titled "Data Format for the Interchange of Fingerprint, Facial & SMT Information" and carries the ANSI designation of ANSI/NIST-ITL 1a-1997.

During a workshop that was held in September 1998, these two standards were reviewed and discussed. Several agreements were reached affecting various aspects of the standard. These included the addition of new fields, new record structures, and the merging of the two standards into this single standard. This standard is the combination of the two previous standards plus additional record type descriptions, the incorporation of agreements reached during the September workshop, and subsequent proposals offered for improving the standard.

1 Scope, purpose, and application

1.1 Scope

This standard defines the content, format, and units of measurement for the exchange of fingerprint, palmprint, facial/mugshot, and scar, mark, & tattoo (SMT) image information that may be used in the identification process of a subject. The information consists of a variety of mandatory and optional items, including scanning parameters, related descriptive and record data, digitized fingerprint information, and compressed or uncompressed images. This information is intended for interchange among criminal justice administrations or organizations that rely on automated fingerprint and palmprint identification systems or use facial/mugshot or SMT data for identification purposes.

This standard does not define the characteristics of the software that shall be required to format the textual information or to compress and assemble the associated digital fingerprint image information. Typical applications for this software might include, but are not limited to, computer systems associated with a live-scan fingerprinting system, a workstation that is connected to or is part of an Automated Fingerprint Identification System (AFIS), or an Image Storage and Retrieval system containing fingerprints, facial/mugshot, or SMT images.

1.2 Purpose

Information compiled and formatted in accordance with this standard can be recorded on machine-readable media or may be transmitted by data communication facilities in lieu of a fingerprint card, a latent fingerprint, facial/mugshot, or other types of photographs. Law enforcement and criminal justice agencies will use the standard to exchange fingerprint, palmprint, or other photographic images and related identification data.

1.3 Application

Systems claiming conformance with this standard shall implement the transmitting and/or receiving of record types as defined by this standard. Systems claiming conformance are not required to implement every record type specified herein. At a minimum, they must be capable of transmitting and receiving Type-1 records. However, for a transaction to be meaningful, there must be at least one additional type of record included. The implementor must document the record types supported in terms of transmitting and/or receiving. Those record types not implemented shall be ignored by the conforming system.

2 Normative references

The following standards contain provisions that, through reference in this text, constitute provisions of this American National Standard. At the time of publication, the editions indicated were valid. All standards are subject to revision, and parties that utilize this American National Standard are encouraged to investigate the possibility of applying the most recent editions of the standards indicated below.

ANSI X3.4-1986 (R1992), Information Systems --- Coded Character Sets ---7-Bit American National Standard Code for Information Interchange (7-Bit ASCII).

ANSI X3.172-1990, Information Systems --- Dictionary for Information Systems.

ANSI/EIA - 538-1988 Facsimile Coding Schemes and Coding Control Functions for Group 4 Facsimile Equipment.

ANSI/IAI 2-1988, Forensic Identification --- Automated Fingerprint Identification Systems --- Glossary of Terms and Acronyms.

CJIS-RS-0010 (V7) Electronic Fingerprint Transmission Specification[1].

IAFIS-IC-0110 (V3) WSQ Gray-scale Fingerprint Image Compression Specification[2].

ISO 646-1983 7-Bit Coded Character Set for Information Interchange.[3]

ISO 8601-1988, Data Elements and Interchange Formats - Information Interchange Representation of Dates and Times.[4]

ANSI/NIST-CSL 1-1993, Information systems – Data Format for the Interchange of Fingerprint Information.

ANSI/NIST-ITL 1a-1997, Information systems – Data Format for the Interchange of Fingerprint, Facial, and SMT Information.

ISO International Standard 10918-1, Information Technology - Digital Compression and Coding of Continuous-Tone Still Images Part 1: Requirements and Guidelines[5]. This is commonly referred to as the JPEG (Joint Photographic Experts Group) algorithm.

National Crime Information Center (NCIC) Code Manual, Eighth Edition, July 14, 1999[6].

---

[1] Available from Criminal Justice Information Services Division, Federal Bureau of Investigation 935 Pennsylvania Avenue, NW, Washington, DC 20535.

[2] Available from Criminal Justice Information Services Division, Federal Bureau of Investigation, 935 Pennsylvania Avenue, NW, Washington, DC 20535.

[3] Available from the American National Standards Institute, 11 West 42nd Street, New York, NY 10036.

[4] Available from the American National Standards Institute, 11 West 42nd Street, New York, NY 10036.

[5] Available from the American National Standards Institute, 11 West 42nd Street, New York, NY 10036.

[6] Available from the U.S. Department of Justice, Federal Bureau of Investigation, 935 Pennsylvania Avenue, NW, Washington, DC 20535.

3 Definitions

The following definitions and those given in the American National Standard Automated Fingerprint Identification Systems --- Glossary of Terms and Acronyms, ANSI/IAI 2-1988, apply to this standard.

3.1 ANSI: Abbreviation for the American National Standards Institute, Inc.

3.2 aspect ratio: The width-to-height ratio of the captured image.

3.3 effective scanning resolution: The number of pixels per unit distance that remain after a captured image has been subsampled, scaled, or interpolated down to produce an image having a lower value of scanning resolution (fewer pixels per mm) than was used originally to capture the image.

3.4 logical record: A record independent of its physical environment; portions of one logical record may be located in different physical records, or several logical records or parts of logical records may be located in one physical record.

3.5 minutia: The point where a friction ridge begins, terminates, or splits into two or more ridges. Minutiae are friction ridge characteristics that are used to individualize a fingerprint image.

3.6 mugshot: Term used interchangeably with facial image. The term facial image usually implies a higher quality image than a mugshot.

3.7 native scanning resolution: The nominal scanning resolution used by a specific AFIS, live-scan reader, or other image capture device and supported by the originator of the transmission.

3.8 nominal transmitting resolution: The nominal number of pixels per unit distance (ppmm or ppi) of the transmitted image. The transmitting resolution may be the same as the scanning resolution for a particular image. On the other hand, the transmitting resolution may be less than the scanning resolution if the scanned image was subsampled, scaled, or interpolated down before transmission.

3.9 ppi: Abbreviation for pixels per inch.

3.10 ppmm: Abbreviation for pixels per millimeter.

3.11 RGB: Red, Green, Blue used to represent color pixels comprised of a specified number of bits to represent each of these primary color components.

3.12 SMT: Abbreviation used for scar, mark, and tattoo information.

3.13 scanning resolution: The number of pixels per unit distance at which an image is captured (ppmm or ppi).

3.14 tagged-field record: A logical record containing unique ASCII field identifiers for variable-length data fields that is capable of being parsed based on the contents of the first two fields.

3.15 transaction: A command, message, or input record that explicitly or implicitly calls for a processing action. Information contained in a transaction shall be applicable to a single subject.

4 Transmitted data conventions

4.1 Byte and bit ordering

Each information item, subfield, field, and logical record shall contain one or more bytes of data. Within a file, the order for transmission of both the ASCII and the binary representations of bytes shall be most significant byte first and least significant byte last. Within a byte, the order of transmission shall be the most significant bit first and the least significant bit last. Figure 1 illustrates the order of transmission of the bytes and bits within a file.

4.2 Grayscale data

Grayscale image data may be transmitted in either compressed or uncompressed form.

The transmission of uncompressed grayscale images shall consist of pixels, each of which shall normally be quantized to eight bits (256 gray levels) and held in a single unsigned byte. Increased precision for pixel values greater than 255 shall use two unsigned bytes to hold sixteen-bit pixels with values in the range of 0-65635. For

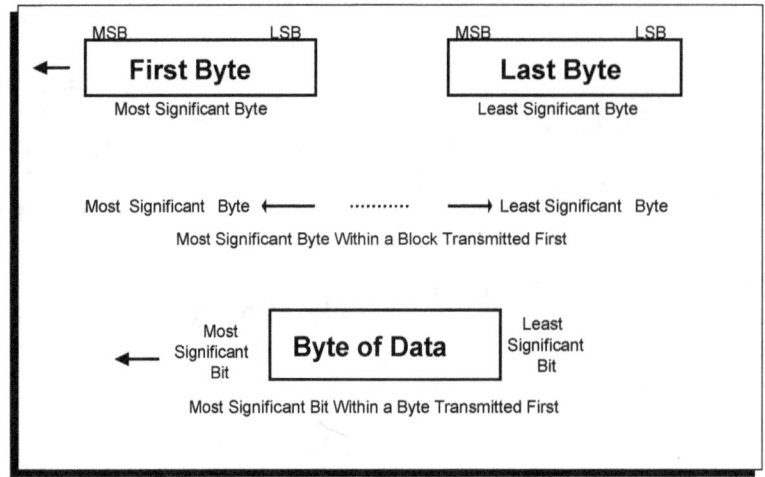

Figure 1 – Byte & bit ordering

grayscale data, a true black pixel shall be represented by a zero. A true white pixel shall have all of its bits of precision set to "1". Therefore, true white pixels quantized to eight bits shall have a value of "255", while a value of "1023" shall be used for pixels quantized to ten bits.

The transmission of compressed grayscale images shall be the output of the appropriate grayscale compression algorithm specified. Upon decompression the grayscale value for each pixel shall be represented in the same manner as pixels in an uncompressed image.

4.3 Binary data

Binary image data may be transmitted in either compressed or uncompressed form.

The transmission of uncompressed binary images shall consist of pixels, each of which shall be quantized to one of two levels (binary representation). A value of zero shall be used to represent a white pixel and a value of one shall be used to represent a black pixel. For transmission of uncompressed binary images, eight pixels shall be left justified and packed into a single unsigned byte. The most significant bit of the byte shall be the first of the eight pixels scanned.

The transmission of compressed binary images shall be the output of the binary compression algorithm specified by ANSI/EIA-538-1988. Upon decompression, each pixel with a value of zero shall be considered to be white and each pixel with a value of one shall be considered to be black.

4.4 Color data

For color facial, SMT, or testing images scanned, it shall be assumed that the scanned images consist of nominal 24-bit RGB pixels.

Color image data may be transmitted in either compressed or uncompressed form. The transmission of uncompressed color images shall consist of RGB pixels, each of which shall be quantized to 256 levels (8 bits) for each of the three components. For each pixel, the three components shall be sequentially formatted for transmission on a pixel-by-pixel basis. Compressed image data shall adhere to the requirements of the algorithm used.

4.5 Scan sequence

For each color, grayscale, or binary image that was captured and formatted, the transmitted scan sequence shall be assumed to have been left to right and top to bottom.

For the purpose of describing the position of each pixel within an image to be exchanged, a pair of reference axes shall be used. The origin of the axes, pixel location (0,0), shall be located at the upper left-hand corner of each image. The x-coordinate (horizontal) position shall increase positively from the origin to the right side of the image. The y-coordinate (vertical) position shall increase positively from the origin to the bottom of the image.

5 Image resolution requirements

Image resolution requirements are applicable to fingerprint and palmprint images. Facial/mugshot and SMT images rely on the total number of pixels scanned and transmitted and are not dependent on the specific scanning resolution used.

5.1 Scanner resolution requirement

Binary and grayscale fingerprint images to be exchanged shall be captured by an AFIS, live-scan reader, or other image capture device operating at a specific native scanning resolution. The minimum scanning resolution for this capture process shall be 19.69 ppmm plus or minus 0.20 ppmm (500 ppi plus or minus 5 ppi). Scanning resolutions greater than this minimum value and with a device tolerance of plus or minus 1% may be used. Although a minimum scanning resolution is specified, a maximum value for scanning resolution is not specified by this Standard.

The recommended migration path to higher scanning resolutions for image capturing devices with a native scanning resolution of 19.69 ppmm (500 ppi) shall be at a rate of 100% of the current native scanning resolution. The recommended migration path progresses from 19.69 ppmm to 39.38 ppmm (500 ppi to 1000 ppi), from 39.38 ppmm to 78.76 ppmm (1000 ppi to 2000 ppi), etc. Capture devices with native scanning resolutions, not in step with this migration path, shall provide (through subsampling, scaling, or interpolating downward) an effective scanning resolution that matches the next lower interval in the migration path. For example a device with native scanning resolution of 47.24 ppmm (1200 ppi) shall be required to provide an effective resolution of 39.38 ppmm (1000 ppi).

5.2 Transmitting resolution requirement

Each image to be exchanged shall have a specific resolution associated with the transmitted data. This transmitting resolution does not have to be the same as the scanning resolution. However, the transmitting resolution shall be within the range of permissible resolution values for that record type. When an image is captured at a scanning resolution greater than the permissible upper limit of the transmitting resolution for that record type, the image shall be subsampled, scaled, or interpolated down. This processing to reduce the scanning resolution to a lower effective resolution must be performed before the transmission occurs.

For high-resolution binary and grayscale images, the preferred transmitting resolution shall be the same as the minimum scanning resolution of 19.69 ppmm plus or minus 0.20 ppmm (500 ppi plus or minus 5 ppi). Any transmitting resolution within the range of the minimum scanning resolution to a value of 20.47 ppmm plus or minus 0.20 ppmm (520 ppi plus or minus 5 ppi) is permitted for the processing of high-resolution records.

For low-resolution binary and grayscale images, the preferred transmitting resolution shall be half of the minimum scanning resolution or 9.84 ppmm plus or minus 0.10 ppmm (250 ppi plus or minus 2.5 ppi). Any transmitting resolution within the range of half of the minimum scanning resolution to a value of 10.24 ppmm plus or minus 0.10 ppmm (260 ppi plus or minus 2.5 ppi) is permitted for the processing of low-resolution records.

For variable-resolution images, the preferred transmitting resolution is not specified, but must be at least as great as the high-resolution rate of 19.69 ppmm. At this time there is no upper limit on the variable-resolution rate for transmission. However, the recommended migration path to higher transmitting resolutions is the same as for the scanning resolutions. That is, to progress from 19.69 ppmm to 39.38 ppmm plus or minus 1% (500 ppi to 1000 ppi), from 39.38 ppmm to 78.76 ppmm plus or minus 1% (1000 ppi to 2000 ppi), etc. For images captured at a native scanning resolution greater than the permissible upper limit of a transmitting resolution step in the migration path, it may be necessary to subsample, scale, or interpolate down. The result of this processing is to obtain an effective scanning resolution that conforms to a step in the transmission migration path.

The transmitting resolution shall be contained in fields specified by the format for the variable-resolution record. However, before transmitting variable-resolution records, the operational capabilities of the sending and receiving systems should be addressed, and prior agreement should be made with the recipient agency or organization before transmitting the image.

Table 1 – Logical record types

Logical record Identifier	Logical record contents	Type of data
1	Transaction Information	ASCII
2	Descriptive Text (User-defined)	ASCII
3	Fingerprint Image Data (Low-resolution grayscale)	Binary
4	Fingerprint Image Data (High-resolution grayscale)	Binary
5	Fingerprint Image Data (Low-resolution binary)	Binary
6	Fingerprint Image Data (High-resolution binary)	Binary
7	Image Data (User-defined)	Binary
8	Signature Image Data	Binary
9	Minutiae Data	ASCII
10	Facial & SMT Image Data	ASCII/Binary
11	Reserved for Future Use	-
12	Reserved for Future Use	-
13	Latent Image Data (Variable-resolution)	ASCII/Binary
14	Tenprint fingerprint Impressions (Variable-resolution)	ASCII/Binary
15	Palmprint Image Data (Variable-resolution)	ASCII/Binary
16	User-defined Testing Image Data (Variable-resolution)	ASCII/Binary

6 File description

This standard defines the composition of a transaction file that is transmitted to a remote site or agency. As specified in this standard, certain portions of the transmission shall be in accordance with definitions provided by the receiving agency. This file shall contain one or more logical records each corresponding to one of the defined available types. The logical records are intended to convey specific types of related information pertinent to the transaction itself or to the subject of the transaction. All of the logical records belonging to a single transaction shall be contained in the same physical file.

The standard defines three logical records to exchange ASCII textual information fields, six logical records to exchange binary information, and five tagged-field record types designed to exchange a combination of ASCII and image data within a single logical record structure. These tagged-field records consist of ASCII tagged textual fields and binary, grayscale, or color image data. At the beginning of the record, a series of tagged-fields shall be used to provide information required for processing the image data present in the last field of the logical record.

Two additional record types are reserved for inclusion in future revisions of this standard. The sixteen types of logical records together with the identifier for each type are listed in Table 1.

Table 2 – Number of logical records per transaction

Type of logical record	Tenprint inquiry	Latent inquiry	File maintenance	Image request	Search response	Image request response
1	1	1	1	1	1	1
2	1-N	1-N	1-N	1	1	1
3	0-14	0	0-14	0	0-14	0-14
4	0-14	0-10	0-14	0	0-14	0-14
5	0-14	0	0-14	0	0-14	0-14
6	0-14	0-10	0-14	0	0-14	0-14
7	0	0-N	0-N	0	0-N	0-N
8	0-2	0	0-2	0	0-2	0-2
9	0-10	0-N	0-N	0	0	0
10	0-N	0-N	0-N	0	0-N	0-N
13	0	0-N	0-N	0	0-N	0-N
14	0-14	0	0-14	0	0-N	0-N
15	0-6	0	0-6	0	0-N	0-N
16	0	0	0-N	0	0-N	0-N

6.1 File format

A file shall contain one or more logical records pertaining to a single subject. The data in the Type-1 record shall always be recorded in variable length fields using the 7-bit American National Standard Code for Information Interchange (ASCII) as described in ANSI X3.4-1986 and Annex A. For purposes of compatibility, the eighth (leftmost) bit shall contain a value of zero.

The text or character data in Type-2, and Type-9 through Type-16 records will normally be recorded using the 7-bit ASCII code in variable-length fields with specified upper limits on the size of the fields. The first two fields in all tagged-field records must be numerically ordered. The remaining textual fields can be in any order. For tagged-field image records, Type-10 through Type-16, the last and concluding field shall have a tagged ASCII field number identifier followed by the image data.

For data interchange between non-English speaking agencies, character sets other than 7-bit ASCII may be used in textual fields contained in Type-2 and Type-9 through Type-16 records. The mechanism to change character sets is described in Section 7.2.3, International Character Sets.

For the binary image Type-3, Type-4, Type-5, Type-6, and Type-8 logical records, the content and order of the recorded fields are specified by this standard. With the exception of the first two fields, the remaining fields of the Type-7 logical image record are all user-defined. All fields and data in these record types shall be recorded as binary information.

6.2 File contents

Files to be exchanged are required to contain one and only one Type-1 logical record per transaction. The Type-1 logical record shall always be the first logical record within the file. Depending on the usage, the number of fingerprint, palmprint, facial/mugshot, or SMT images available for processing, one or more additional records may be present in the file.

Table 2 lists the typical range or the number of records that may be contained in a file. These record counts are shown by logical record types for common processing functions used for search inquiries, file maintenance, image request, and image responses. The record limits stated in the table are examples of typical transactions and should only be interpreted as a guideline. In particular, specific applications must dictate the maximum number of Type-2 records that may be present in a transaction. The ranges listed specify the minimum and maximum number of logical records that may be contained in the file. The mandatory inclusion of a logical record is indicated by an entry of "1" in the table. An entry of "0" indicates the exclusion of that logical record type. The appearance of "0-N" in the table indicates that the standard imposes no limits on the number of records for that logical record type. An entry of "1-N" requires that at least one record be present with no upper limit on the number of records that may be present. However, any recipient agency may impose their own specific limit for each type of logical record.

6.3 Implementation domains

The Type-2 record is composed of user-defined textual fields. Many of the information fields in the Type-2 record are used in the same way by local, state, and Federal agencies and require the same data and formatting. In order to establish a common basis for field numbering, meaning, and formatting, jurisdictions that use the same general set of data fields may subscribe to a common implementation domain.

An implementation domain can be viewed as a group of agencies or organizations that have agreed to use specific pre-assigned groups of numbered tagged fields for exchanging information unique to their installations. Each tagged-field number shall also have a definition and format associated with it. The domain implementation name uniquely identifies field contents and avoids tag numbers with multiple Type-2 field definitions. Each domain created shall have a point of contact responsible for keeping the list of numbered tagged fields and assigning new numbered tagged fields to organizations within their domain. The contact shall serve as a registrar and maintain a repository including documentation for all the common and user-specific Type-2 fields contained in the implementation. As additional fields are required by specific agencies for their own applications, new field tag numbers and definitions can be registered and reserved to have a specific meaning. When this occurs, the domain registrar is responsible for registering a single definition for each tagged-field number used by different members of the domain.

The Federal Bureau of Investigation (FBI) has established and maintains the North American Domain subscribed to by the Royal Canadian Mounted Police (RCMP), the FBI, and several state and Federal agencies in North America. The registrar for this domain assigns and accounts for a set of tagged fields to be used by its clients during the processing of transactions. Other domains also exist including those maintained by the United Kingdom (UK) and Interpol. These organizations have developed their own Type-2 record implementations tailored to their specific communities.

6.4 Image designation character (IDC)

With the exception of the Type-1 logical record, each of the remaining logical records present in a file shall include a separate field containing the Image Designation Character (IDC). The IDC shall be used to relate information items in the file contents field of the Type-1 record to each logical record, and to properly identify and link together logical records that pertain to the same subject matter. The value of the IDC shall be a sequentially assigned positive integer starting from zero and incremented by one. If two or more logical records that are different representations of the same subject matter are present in a file, each of those logical records shall contain the same IDC. For example, a high-resolution image record of a specific fingerprint and the corresponding minutiae record for the same finger would carry the same IDC number.

Although there is no upper limit on the number of logical records that may be present in a file, generally a minimum of two, and no more than 25 logical records will be present in a file. A tenprint search inquiry transaction may consist of a Type-1 record, a Type-2 record, 14 high-resolution Type-4 or variable-resolution Type-14 grayscale image records, two Type-8, six palmprint records, and a facial/mugshot image of the subject. Additional mugshot and SMT logical records may expand the file even more. For this file configuration, the IDC shall range from "0" to "23" which would include an IDC code for the Type-2 record. Within the same file, multiple logical record types may be present and represent the same image. For example, if core and delta location information for the rolled impressions is requested, the transmission may also need to accommodate ten minutiae records within the same file. For each image representing the ten finger positions, the same IDC would be used in both the image and minutiae records.

The IDC shall also be used to relate information items in the file contents field of the Type-1 record to each facial or SMT image record. It properly identifies and links together different logical record types created from the same face/mugshot or SMT image.

Furthermore, zero or more Type-7 records may also be present. Each Type-7 logical record representing a specific subject matter shall have a unique IDC with an increment of one greater than the last IDC used.

7 Record description

7.1 Logical record types

7.1.1 Type-1 transaction record

A Type-1 logical record is mandatory and shall be contained in each transaction. The Type-1 record shall provide information describing type and use or purpose for the transaction involved, a listing of each logical record included in the file, the originator or source of the physical record, and other useful and required information items.

7.1.2 Type-2 user-defined descriptive text record

Type-2 logical records shall contain user-defined textual fields providing identification and descriptive information about the subject of the fingerprint information. Data contained in this record shall conform in format and content to the specifications of the domain name as listed in Domain Name field found in the Type-1 record.

7.1.3 Type-3 low-resolution grayscale record

Type-3 logical records shall contain, and be used to exchange low-resolution grayscale fingerprint image data that was scanned at no less than the minimum scanning resolution and then subsampled, scaled down, or interpolated. The resultant transmitting resolution shall be within the bounds of the permissible transmitting resolutions for low-resolution fingerprint images.

The low-resolution grayscale fingerprint image data contained in the Type-3 logical record may be in compressed form. Typically, there may be up to 14 of these Type-3 records in a file; ten rolled impressions of the individual fingers, two plain impressions of the thumbs, and two simultaneously obtained plain impressions of the four remaining fingers on each hand.

7.1.4 Type-4 high-resolution grayscale record

Type-4 logical records shall contain, and be used to exchange high-resolution grayscale fingerprint image data that was scanned at no less than the minimum scanning resolution. If the scanning resolution is greater than the upper limit of the permissible transmitting resolution, the scanned data shall be subsampled, scaled down, or interpolated. The resultant transmitting resolution shall be within the bounds of the permissible transmitting resolutions for high-resolution fingerprint images.

The high-resolution grayscale fingerprint image data contained in the Type-4 logical record may be in compressed form. Typically, there may be up to 14 of these Type-4 records in a file; ten rolled impressions of the individual fingers, two plain impressions of the thumbs, and two simultaneously obtained plain impressions of the four remaining fingers on each hand.

7.1.5 Type-5 low-resolution binary record

Type-5 logical records shall contain, and be used to exchange low-resolution binary fingerprint image data that was scanned at no less than the minimum scanning resolution and then subsampled, scaled down, or interpolated. The resultant transmitting resolution shall be within the bounds of the permissible transmitting resolutions for low-resolution fingerprint images.

The low-resolution binary fingerprint image data contained in the Type-5 logical record may be in compressed form. Typically, there may be up to 14 of these Type-5 records in a file; ten rolled impressions of the individual fingers, two plain impressions of the thumbs, and two simultaneously obtained plain impressions of the four remaining fingers on each hand.

7.1.6 Type-6 high-resolution binary record

Type-6 logical records shall contain, and be used to exchange high-resolution binary fingerprint image data that was scanned at no less than the minimum scanning resolution. If the scanning resolution is greater than the upper limit of the permissible transmitting resolution, the scanned data shall be subsampled, scaled down, or interpolated. The resultant transmitting resolution shall be within the bounds of the permissible transmitting resolutions for high-resolution fingerprint images.

The high-resolution binary fingerprint image data contained in the Type-6 logical record may be in compressed form. Typically, there may be up to 14 of these Type-6 records in a file; ten rolled impressions of the individual fingers, two plain impressions of the thumbs, and two simultaneously obtained plain impressions of the four remaining fingers on each hand.

7.1.7 Type-7 user-defined image record

Type-7 logical records shall contain user-defined image data. This record type is used to exchange image data that is not elsewhere specified or described in this standard. With the exception of the length and IDC fields, the format, parameters, and types of images to be exchanged are undefined by this standard and shall be agreed upon between the sender and recipient. This record type is included to handle miscellaneous images such as those pertaining to wrists, toes, soles, etc.

7.1.8 Type-8 signature image data record

Type-8 logical records shall contain, and be used to exchange scanned high-resolution binary or vectored signature image data. If scanned, the resolution of the image data shall be no less than the minimum scanning resolution. If necessary, the scanned image data shall be subsampled, scaled down, or interpolated to fall within the limits of the transmitting resolution requirement. The resultant transmitting resolution shall be within the bounds of the permissible transmitting resolutions for the high-resolution fingerprint images. Vectored signature data shall be expressed as a series of binary numbers.

Typically, there may be up to two of these Type-8 signature records in a file. Each Type-8 record shall contain image data representing the signature of the person being fingerprinted or of the official taking the fingerprint.

7.1.9 Type-9 minutiae record

Type-9 logical records shall contain, and be used to exchange geometric and topological minutiae and related information encoded from a finger or palm. Each record shall represent the processed image data from which the location and orientation descriptors of extracted minutiae characteristics are listed. The primary use of this record type shall be for remote searching of latent prints. Each Type-9 logical record shall contain the minutiae data read from a fingerprint, palm, or latent image.

7.1.10 Type-10 facial & SMT image record

Type-10 tagged-field image records shall contain, and be used to exchange facial and SMT image data together with textual information pertinent to the digitized image. The source of the image data shall be the image captured from scanning a photograph, a live image captured with a digital camera, or a digitized "freeze-frame" from a video camera.

7.1.11 Type-11 record reserved for future use

7.1.12 Type-12 record reserved for future use

7.1.13 Type-13 variable-resolution latent image record

Type-13 tagged-field image records shall contain, and be used to exchange variable-resolution latent fingerprint and palmprint image data together with fixed and user-defined textual information fields pertinent to the digitized image. It is strongly recommended that the minimum scanning resolution (or effective scanning resolution) and transmission rate for latent images be 39.38 ppmm plus or minus 0.40 ppmm (1000 ppi plus or minus 10 ppi). However, in all cases the scanning resolution used to capture a latent image shall be at least as great as the minimum scanning resolution of 19.69 ppmm (500 ppi).

The variable-resolution latent image data contained in the Type-13 logical record shall be uncompressed or may be the output from a lossless compression algorithm. There is no limit on the number of these latent records that may be present in a transaction.

7.1.14 Type-14 variable-resolution tenprint image record

Type-14 tagged-field image records shall contain, and be used to exchange variable-resolution tenprint fingerprint image data together with fixed and user-defined textual information fields pertinent to the digitized image. Fingerprint images can be either rolled or plain impressions.

The scanning resolution is not specified for this record type. While the Type-14 record may be used for the exchange of 19.69 ppmm (500 ppi) images, it is strongly recommended that the minimum scanning resolution (or effective scanning resolution) for tenprint images be 39.38 ppmm plus or minus 0.40 ppmm (1000 ppi plus or minus 10 ppi). It should be noted that as the resolution is increased, more detailed ridge and structure information becomes available in the image. However, in all cases the scanning resolution used to capture a tenprint image shall be at least as great as the minimum scanning resolution of 19.69 ppmm (500ppi).

The variable-resolution tenprint image data contained in the Type-14 logical record may be in a compressed form. Typically, there may be up to 14 of these Type-14 records in a file; ten rolled impressions of the individual fingers, two plain impressions of the thumbs and two plain impressions of the four simultaneously obtained remaining fingers of each hand.

7.1.15 Type-15 variable-resolution palmprint image record

Type-15 tagged-field image records shall contain, and be used to exchange variable-resolution palmprint image data together with fixed and user-defined textual information fields pertinent to the digitized image. Image data contained in this record may be the full palm impression, the upper half of the palm, the lower half of the palm, or the writer's palmprint impression.

The scanning resolution is not specified for this record type. While the Type-15 record may be used for the exchange of 19.69 ppmm (500 ppi) images, it is strongly recommended that the minimum scanning resolution (or effective scanning resolution) for palmprint images be 39.38 ppmm plus or minus 0.40 ppmm (1000 ppi plus or minus 10 ppi). It should be noted that as the resolution is increased, more detailed ridge and structure information becomes available in the image. However, in all cases the scanning resolution used to capture a tenprint image shall be at least as great as the minimum scanning resolution of 19.69 ppmm (500ppi).

The variable-resolution palmprint image data contained in the Type-15 logical record may be in a compressed form. Typically, there may be up to 6 of these Type-15 records in a file; two full palmprints or four partial palms, and two writer's palms.

7.1.16 Type-16 user-defined testing image record

The Type-16 tagged-field image record is intended as the tagged-field version of the Type-7 user-defined logical record. It is designed for developmental purposes and for the exchange of miscellaneous images. This tagged-field logical record shall contain, and be used to exchange, image data together with textual information fields pertinent to the digitized image. Such an image is usually not elsewhere specified or described in this Standard.

The scanning resolution is not specified for this record type but shall be at least as great as the minimum scanning resolution, that is, 19.69 ppmm (500ppi). Increases in the resolution used for capturing images should follow the recommended migration path to 39.38 ppmm (1000 ppi), 78.76 ppmm (2000 ppi), etc. It should be noted that as the resolution is increased, more detailed ridge and structure information becomes available in the image.

The variable-resolution image data contained in the Type-16 logical record may be in a compressed form. With the exception of the first two tagged fields and the descriptors for the image data, the remaining details of the Type-16 record are undefined by this standard and shall be agreed upon between the sender and recipient.

7.2 Record format

A transaction file shall consist of one or more logical records. For each logical record contained in the file, several information fields appropriate to that record type shall be present. Each information field may contain one or more basic single-valued information items. Taken together these items are used to convey different aspects of the data contained in that field. An information field may also consist of one or more information items grouped together and repeated multiple times within a field. Such a group of information items is known as a subfield. An information field may therefore consist of one or more subfields of information items.

Table 3 – Information separators

ASCII character	Column / row position	Description
FS	1 / 12	Separates logical records of a file or is the terminating character of a transaction
GS	1 / 13	Separates fields of a logical record
RS	1 / 14	Separates multiple data entries (subfields) of an Information field
US	1 / 15	Separates individual information items of the field or subfield

7.2.1 Information separators

In the tagged-field logical records (Type-1, Type-2, and Type-9 through Type-16), mechanisms for delimiting information are implemented by use of the four ASCII information separators. The delimited information may be items within a field or subfield, fields within a logical record, or multiple occurrences of subfields. These information separators are defined in the referenced standard ANSI X3.4 whose code table is shown in Annex A. These characters are used to separate and qualify information in a logical sense. Viewed in a hierarchical relationship, the File Separator "FS" character is the most inclusive followed by the Group Separator "GS", the Record Separator "RS", and finally the Unit Separator "US" characters. Table 3 lists these ASCII separators, the column/row position in the ASCII table shown in Annex A, and a description of their use within this standard.

Information separators should be functionally viewed as an indication of the type data that follows. The "US" character shall separate individual information items within a field or subfield. This is a signal that the next information item is a piece of data for that field or subfield. Multiple subfields within a field separated by the "RS" character signals the start of the next group of repeated information item(s). The "GS" separator character used between information fields signals the beginning of a new field preceding the field identifying number that shall appear. Similarly, the beginning of a new logical record shall be signaled by the appearance of the "FS" character.

These separators shall be in addition to any other symbols, punctuation, or delimiters as specified in this standard. Annex B illustrates the use of these characters, and Annex F provides examples of their use of the standard.

The four characters are only meaningful when used as separators of data items in the fields of the ASCII text records. There is no specific meaning attached to these characters occurring in binary image records and binary fields – they are just part of the exchanged data.

Normally, there should be no empty fields or information items and therefore only one separator character should appear between any two data items. The exception to this rule occurs for those instances where the data in fields or information items in a transaction are unavailable, missing, or optional, and the processing of the transaction is not dependent upon the presence of that particular data. In those instances, multiple and adjacent separator characters shall appear together rather than requiring the insertion of dummy data between separator characters.

Consider the definition of a field that consists of three information items. If the information for the second information item is missing, then two adjacent "US" information separator characters would occur between the first and third information items. If the second and third information items were both missing, then three separator characters should be used – two "US" characters in addition to the terminating field or subfield separator character. In general, if one or more mandatory or optional information items are unavailable for a field or subfield, then the appropriate number of separator character should be inserted.

It is possible to have side-by-side combinations of two or more of the four available separator characters. When data are missing or unavailable for information items, subfields, or fields, there must be one fewer separator characters present than the number of data items, subfields, or fields required.

7.2.2 Record layout

For tagged-field logical records (Type-1, Type-2, Type-9, Type-10, and Type-13 through Type-16), each information field that is used shall be numbered in accordance with this standard. The format for each field shall consist of the logical record type number (chosen from Table-1) followed by a period ".", a field number followed by a colon ":", followed by the information appropriate to that field. The tagged-field number can be any one- to nine-digit number occurring between the period "." and the colon ":". It shall be interpreted as an unsigned integer field number. This implies that a field number of "2.123:" is equivalent to and shall be interpreted in the same manner as a field number of "2.000000123:".

NOTE: For purposes of illustration throughout this document, a three-digit number shall be used for enumerating the fields contained in each of the tagged-field logical records described herein. Field numbers will have the form of "TT.xxx:" where the "TT" represents the one- or two-character record type followed by a period. The next three characters comprise the appropriate field number followed by a colon. Descriptive ASCII information or the image data follows the colon.

Logical Type-1, Type-2, and Type-9 records contain only ASCII textual data fields. The entire length of the record (including field numbers, colons, and separator characters) shall be recorded as the first ASCII field within each of these record types. The ASCII File Separator *"FS"* control character (signifying the end of the logical record or transaction) shall follow the last byte of ASCII information and shall be included in the length of the record.

In contrast to the tagged-field concept, the Type-3 through Type-8 records contain only binary data recorded as ordered fixed-length binary fields. The entire length of the record shall be recorded in the first four-byte binary field of each record. For these binary records, neither the record number with its period, nor the field identifier number and its following colon, shall be recorded. Furthermore, as all the field lengths of these six records are either fixed or specified, none of the four separator characters (*"US"*, *"RS"*, *"GS"*, or *"FS"*) shall be interpreted as anything other than binary data. For these binary records, the *"FS"* character shall not be used as a record separator or transaction terminating character.

The Type-10 and Type-13 through Type-16 tagged-field image records combine ASCII fields with a single binary image field. Each ASCII field contains a numeric field identifier and its descriptive data. The last physical field in a tagged-field image record shall always be numbered 999 and shall contain the image data placed immediately following the colon (":") of the field identifier. The record length field shall contain the length of the record. The ASCII File Separator *"FS"* control character shall follow the last byte of the compressed or uncompressed image data. The *"FS"* character shall signify the end of the logical record or transaction and shall be included as part of the record length.

7.2.3 International character sets

All of the fields in the Type-1 transaction record must be recorded using the 7-bit ASCII code, which is the default character set code within a transaction. In order to effect data and transaction interchanges between non-English based agencies, a technique is available to encode information using character sets other than 7-bit ASCII. Fields from the Type-1 logical record and ASCII "LEN" and "IDC" text fields must still be encoded using 7-bit ASCII. But all other designated text fields can be encoded using alternate character sets. The general mechanism for accomplishing this provides for backward compatibility with existing readers, supports multiple character sets in a single text string, and handles internationally accepted character sets and text order conventions such as ISO character sets and Unicode.

To switch character sets within a transaction, the Type-1 record shall contain a field listing the Directory of Character Sets (DCS) used in the transaction. The DCS is an ordered list of 3 information items containing an identifying code, the name of an international character set, and its version. The code for a specific character set and other special codes shall be embedded in the transaction to signal the conversion to a different international character set. The ASCII Start-of-Text "STX" character (0x02) followed by the equal sign "=" is used to signal the change to an alternate character set defined by the specific DCS code that follows. The entire Start-of-Text sequence is terminated by a single instance of the ASCII End-of-Text "ETX" character (0x03). This alternate character set will remain active until a closing "ETX" character is encountered or the next ASCII information separator character is encountered.

The base-64 encoding scheme, found in email, shall be used for converting non-ASCII text into ASCII form. Annex C describes the use of the base-64 system. By convention, any language or character set text string following the Start-of-Text character sequence will be base-64 encoded for subsequent processing.

The field number including the period and colon, for example "2.001:", in addition to the *"US", "RS", "GS"*, and *"FS"* information separators shall appear in the transaction as 7-bit ASCII characters without conversion to base-64 encoding.

All text between the STX sequence and the closing ETX character shall be encoded in base-64 notation. This is true even when the 7-bit ASCII character set is specified.

8 Type-1 transaction information record

8.1 Fields for Type-1 transaction information record

The following paragraphs describe the data contained in fields for the Type-1 logical record. Each field shall begin with the number of the record type followed by a period followed by the appropriate field number followed by a colon. Annex F contains an example of the use of the standard that illustrates the layout for a Type-1 logical record.

8.1.1 Field 1.001: Logical record length (LEN)

This mandatory ASCII field shall contain the total count of the number of bytes in this Type-1 logical record. Field 1.001 shall begin with "1.001:", followed by the length of the record including every character of every field contained in the record and the information separators. The *"GS"* character shall separate the length code of Field 1.001 from the next field.

NOTE: Although it will not always be explicitly repeated in the remainder of this standard, use of separators within the Type-1, Type-2, and Type-9 through Type-16 logical records shall always be observed. The *"US"* separator shall separate multiple information items within a field or subfield, the *"RS"* separator shall separate multiple subfields, and the *"GS"* separator shall separate information fields.

8.1.2 Field 1.002: Version number (VER)

This mandatory four-byte ASCII field shall be used to specify the current version number of the standard implemented by the software or system creating the file. The format of this field shall consist of four numeric characters. The first two characters shall specify the major version number. The last two characters shall be used to specify the minor revision number. The initial revision number for a version shall be "00". The entry in this field for this 2000 approved standard shall be "0300". This version number signifies the inclusion of the tagged-field logical Type-10 through Type-16 image records.

8.1.3 Field 1.003: File content (CNT)

This mandatory field shall list and identify each of the logical records in the file by record type. It also specifies the order in which the remaining logical records shall appear in the file. It shall consist of two or more subfields. Each subfield shall contain two information items describing a single logical record found in the current file. The subfields shall be entered in the same order in which the logical records shall appear and be transmitted. The *"RS"* separator character shall be entered between the subfields.

The first subfield shall relate to this Type-1 Transaction record. The first information item within this subfield shall be the single character indicating that this is a Type-1 record consisting of header information (the logical record identifier "1" selected from Table-1).

The second information item of this subfield shall be the sum of the Type-2 through Type-16 logical records contained in this file. This number is also equal to the count of the remaining subfields of Field 1.003. The *"US"* separator character shall be entered between the first and second information items.

Each of the remaining subfields of Field 1.003 relate to a single Type-2 through Type-16 logical record contained in the file. Two information items shall comprise each subfield. The first information item shall be the record identifier character(s) chosen from Table-1 that identifies the record type. The second item shall be the IDC associated with the logical record pertaining to that subfield. The IDC shall be a positive integer equal to or greater than zero. The *"US"* character shall be used to separate the two information items.

8.1.4 Field 1.004: Type of transaction (TOT)

This mandatory field shall contain an identifier, which designates the type of transaction and subsequent processing that this file should be given. (Note: Type of Transaction shall be in accordance with definitions provided by the receiving agency.) The last character of this field shall be a *"GS"* separator character used to separate Field 1.004 from the next field.

8.1.5 Field 1.005: Date (DAT)

This mandatory field shall contain the date that the transaction was initiated. The date shall appear as eight digits in the format CCYYMMDD. The CCYY characters shall represent the year of the transaction; the MM characters shall be the tens and units values of the month; and the DD characters shall be the day in the month. For example, "20000103" represents January 3, 2000.

8.1.6 Field 1.006: Priority (PRY)

When this field is used, it shall contain a single information character to designate the urgency with which a response is desired. The values shall range from "1" to "9", with "1" denoting the highest priority. The default value shall be defined by the agency receiving the transaction.

8.1.7 Field 1.007: Destination agency identifier (DAI)

This mandatory field shall contain the identifier of the administration or organization designated to receive the transmission. The size and data content of this field shall be user-defined and in accordance with the receiving agency.

8.1.8 Field 1.008: Originating agency identifier (ORI)

This mandatory field shall contain the identifier of the administration or organization originating the transaction. The size and data content of this field shall be user-defined and in accordance with the receiving agency.

8.1.9 Field 1.009: Transaction control number (TCN)

This mandatory field shall contain the Transaction Control Number as assigned by the originating agency. A unique alphanumeric control number shall be assigned to each transaction. For any transaction that requires a response, the respondent shall refer to this number in communicating with the originating agency.

8.1.10 Field 1.010: Transaction control reference (TCR)

This optional field shall be used for responses that refer to the TCN of a previous transaction involving an inquiry or other action that required a response.

8.1.11 Field 1.011: Native scanning resolution (NSR)

This mandatory field shall specify the native scanning resolution of the AFIS or other fingerprint or palmprint image capture device supported by the originator of the transmission. This field permits the recipient of this transaction to send response data at a transmitting resolution tailored to the NSR (if it is able to do so) or to the minimum scanning resolution. This field shall contain five bytes specifying the native scanning resolution in pixels per millimeter. The resolution shall be expressed as two numeric characters followed by a decimal point and two more numeric characters (e.g., 19.69). This field is needed because the interchange of fingerprint information between systems of the same manufacturer may, in some instances, be more efficiently done at a transmitting resolution equal to the native scanning resolution of the system rather than at the minimum scanning resolution specified in this standard. For applications other than fingerprint where resolution is not a factor or not applicable (such a facial or SMT image) this field shall be set to "00.00".

Table 4 – Directory of character sets

Character set index	Character set name	Description
000	ASCII	7-bit English (Default)
001	ASCII	8-bit Latin
002	UNICODE	16-bit
003-127	-------------	Reserved for ANSI/NIST future use
128-999	-------------	User-defined character sets

8.1.12 Field 1.012: Nominal transmitting resolution (NTR)

This mandatory field shall specify the nominal transmitting resolution for the fingerprint or palmprint image(s) being exchanged. This field shall contain five bytes specifying the transmitting resolution in pixels per millimeter. The resolution shall be expressed as two numeric characters followed by a decimal point and two more numeric characters (e.g., 19.69). The transmitting resolution shall be within the range specified by the transmitting resolution requirement. For applications where resolution is not a factor or not applicable (such as a facial or SMT image) this field shall be set to "00.00".

8.1.13 Field 1.013: Domain name (DOM)

This optional field identifies the domain name for the user-defined Type-2 logical record implementation. If present, the domain name may only appear once within a transaction. It shall consist of one or two information items. The first information item will uniquely identify the agency, entity, or implementation used for formatting the tagged fields in the Type-2 record. An optional second information item will contain the unique version of the particular implementation. The default value for the field shall be the North American Domain implementation and shall appear as "1.013:NORAM{US}{GS}".

8.1.14 Field 1.014: Greenwich mean time (GMT)

This optional field provides a mechanism for expressing the date and time in terms of universal Greenwich Mean Time (GMT) units. If used, the GMT field contains the universal date that will be in addition to the local date contained in Field 1.005 (DAT). Use of the GMT field eliminates local time inconsistencies encountered when a transaction and its response are transmitted between two places separated by several time zones. The GMT provides a universal date and 24-hour clock time independent of time zones. It is represented as "CCYYMMDDHHMMSSZ", a 15-character string that is the concatenation of the date with the GMT and concludes with a "Z". The "CCYY" characters shall represent the year of the transaction, the "MM" characters shall be the tens and units values of the month, and the "DD" characters shall be the tens and units values of the day of the month, the "HH" characters represent the hour, the "MM" the minute, and the "SS" represents the second. The complete date shall not exceed the current date.

8.1.15 Field 1.015: Directory of character sets (DCS)

This optional field is a directory or list of character sets other than 7-bit ASCII that may appear within this transaction. This field shall contain one or more subfields, each with three information items. The first information item is the three-character identifier for the character set index number that references an associated character set throughout the transaction file. The second information item shall be the common name for the character set associated with that index number, the optional third information item is the specific version of the character set used. Table 4 lists the reserved named character sets and their associated 3-character index numbers. The *"US"* character shall separate the first information item from the second and the second from the third. The *"RS"* separator character shall be used between the subfields.

8.2 End of transaction information record Type-1

Immediately following the last information field in the Type-1 logical record, an *"FS"* separator character shall be used to separate it from the next logical record. This *"FS"* character shall replace the *"GS"* character that is normally used between information fields.

9 Type-2 user-defined descriptive text record

Type-2 logical records shall contain textual information relating to the subject of the transaction and shall be represented in an ASCII format. This record may include such information as the state or FBI numbers, physical characteristics, demographic data, and the subject's criminal history. Every transaction shall usually contain one or more Type-2 records which is dependent upon the entry in the Type-of-Transaction Field 1.004 (TOT).

9.1 Fields for Type-2 logical records

The first two data fields of the Type-2 record are mandatory, ordered, and defined by this standard. The remaining fields of the record(s) shall conform to the format, content, and requirements of the subscribed Domain Name (DOM) used by the agency to which the transmission is being sent.

9.1.1 Field 2.001: Logical record length (LEN)

This mandatory ASCII field shall contain the length of the logical record specifying the total number of bytes, including every character of every field contained in the record.

9.1.2 Field 2.002: Image designation character (IDC)

This mandatory field shall be used to identify the user-defined text information contained in this record. The IDC contained in this field shall be the IDC of the Type-2 logical record as found in the file content (CNT) field of the Type-1 record.

9.1.3 Field 2.003-999: User-defined fields

Individual fields required for given transaction types, including field size and content, shall conform to the specifications set forth by the agency to whom the transmission is being sent. Each one to nine digit tagged-field number used in the Type-2 record and its format shall conform to the requirements contained in Section 7.2.2 Record Layout.

9.2 End of Type-2 user-defined descriptive text record

Immediately following the last information field in every Type-2 logical record, an *"FS"* separator shall be used to separate it from the next logical record. This *"FS"* character shall replace the *"GS"* character that is normally used between information fields.

10 Type-3 low-resolution grayscale fingerprint image record

Type-3 logical records shall contain low-resolution grayscale fingerprint image data. The fingerprint image data shall have been scanned at no less than the minimum scanning resolution and then subsampled, scaled down, or interpolated. Alternatively, provided that it is no less than the minimum scanning resolution, the native scanning resolution may be used and the image processed such that the resulting transmitting resolution is within the range specified by the transmitting resolution requirement for low-resolution images. When the image data is obtained from a live-scan reader, it shall be the grayscale subsampled, scaled down, or interpolated output of the live-scan fingerprint scanner and not a rescan of a hard copy fingerprint image.

10.1 Fields for Type-3 logical record

Within a Type-3 logical record, entries shall be provided in nine ordered and unnumbered fields. The first eight fields are fixed length and total eighteen bytes. These fields precede the image data contained in field nine. The size of the ninth field is eighteen bytes less than the value specified in the LEN field.

10.1.1 Logical record length (LEN)

This mandatory four-byte binary field shall occupy bytes one through four. It shall contain the length of the logical record specifying the total number of bytes, including every byte of all nine fields contained in the record.

10.1.2 Image designation character (IDC)

This mandatory one-byte binary field shall occupy the fifth byte of a Type-3 record. It shall be used to identify the image data contained in this record. The IDC contained in this field shall be a binary representation of the IDC found in the file content (CNT) field of the Type-1 record.

10.1.3 Impression type (IMP)

This mandatory one-byte binary field shall occupy the sixth byte of a Type-3 record. The code selected from Table 5, describing the manner by which the fingerprint image information was obtained, shall be entered in this field.

10.1.4 Finger position (FGP)

This mandatory fixed-length field of six binary bytes shall occupy the seventh through twelfth byte positions of a Type-3 record. It shall contain possible finger positions beginning in the leftmost byte of the field (byte seven of the record). The decimal code number corresponding to the known or most probable finger position shall be taken from Table 6 and entered as a binary number right justified and left zero filled within the eight-bit byte. Table 6 also lists the maximum image area and the recommended width and height dimensions for each of the finger positions. The maximum areas are larger than the product of the width and height in order to accommodate images that extend beyond the boundaries of the marked areas on a standard fingerprint card. Up to five additional finger positions may be referenced by entering the alternate finger positions in the remaining five bytes using the same format.

If fewer than five finger position references are to be used, the unused bytes shall be filled with the binary equivalent of "255". The code "0", for "Unknown Finger", shall be used to reference every finger position from one through ten.

10.1.5 Image scanning resolution (ISR)

This mandatory one-byte binary field shall occupy the thirteenth byte of a Type-3 record. It shall contain a binary value of "0" if half the minimum scanning resolution is used, and a "1" if half the native scanning resolution is used.

10.1.6 Horizontal line length (HLL)

This mandatory two-byte binary field shall occupy the fourteenth and fifteenth bytes of the Type-3 record. It shall be used to specify the number of pixels contained on a single horizontal line of the transmitted image.

10.1.7 Vertical line length (VLL)

This mandatory two-byte binary field shall occupy the sixteenth and seventeenth bytes of the Type-3 record. It shall be used to specify the number of horizontal lines contained in the transmitted image.

Table 5 – Finger impression type

Description	Code
Live-scan plain	0
Live-scan rolled	1
Nonlive-scan plain	2
Nonlive-scan rolled	3
Latent impression	4
Latent tracing	5
Latent photo	6
Latent lift	7

10.1.8 Grayscale compression algorithm (GCA)

This mandatory one-byte binary field shall occupy the eighteenth byte of a Type-3 record. It shall be used to specify the type of grayscale compression algorithm used (if any). A binary zero denotes no compression. Otherwise, the contents of this byte shall be a binary representation for the number allocated to the particular compression technique used by the interchange parties. The domain registrar will maintain a registry relating these numbers to the compression algorithms.

10.1.9 Image data

This binary field shall contain all of the low-resolution grayscale image data. Each pixel of the uncompressed image shall be quantized to eight bits (256 gray levels) contained in a single byte. If compression is used, the pixel data shall be compressed in accordance with the compression technique specified in the GCA field. This completes the low-resolution image description for a single image.

Table 6 – Finger position code & maximum size

Finger position	Finger code	Max image area (mm²)	Width (mm)	(in)	Length (mm)	(in)
Unknown	0	1745	40.6	1.6	38.1	1.5
Right thumb	1	1745	40.6	1.6	38.1	1.5
Right index finger	2	1640	40.6	1.6	38.1	1.5
Right middle finger	3	1640	40.6	1.6	38.1	1.5
Right ring finger	4	1640	40.6	1.6	38.1	1.5
Right little finger	5	1640	40.6	1.6	38.1	1.5
Left thumb	6	1745	40.6	1.6	38.1	1.5
Left index finger	7	1640	40.6	1.6	38.1	1.5
Left middle finger	8	1640	40.6	1.6	38.1	1.5
Left ring finger	9	1640	40.6	1.6	38.1	1.5
Left little finger	10	1640	40.6	1.6	38.1	1.5
Plain right thumb	11	2400	25.4	1.0	50.8	2.0
Plain left thumb	12	2400	25.4	1.0	50.8	2.0
Plain right four fingers	13	6800	81.3	3.2	50.8	2.0
Plain left four fingers	14	6800	81.3	3.2	50.8	2.0

10.2 End of Type-3 low-resolution grayscale fingerprint image record

Since the Type-3 logical record is a defined and specified series of binary data fields, no additional bytes shall be transmitted to signify the end of this logical record type.

10.3 Additional low-resolution grayscale fingerprint image records

Typically, up to thirteen more images may be described within the file. For each additional image, a Type-3 logical record is required.

11 Type-4 high-resolution grayscale fingerprint image record

Type-4 logical records shall contain high-resolution grayscale fingerprint image data that have been scanned at the minimum scanning resolution. Alternatively, the native scanning resolution may be used and the image processed such that the resulting transmitting resolution is within the range specified by the transmitting resolution requirement for high-resolution images. When the image data is obtained from a live-scan reader it shall be the grayscale output of the live-scan fingerprint scanner and not a rescan of a hard copy fingerprint image.

11.1 Fields for Type-4 logical record

Within a Type-4 logical record, entries shall be provided in nine ordered and unnumbered fields. The first eight fields are fixed length and total eighteen bytes. These fields precede the image data contained in field nine. The size of the ninth field is eighteen bytes less than the value specified in the LEN field.

11.1.1 Logical record length (LEN)

This mandatory four-byte binary field shall occupy bytes one through four. It shall contain the length of the logical record specifying the total number of bytes, including every byte of all nine fields contained in the record.

11.1.2 Image designation character (IDC)

This mandatory one-byte binary field shall occupy the fifth byte of a Type-4 record. It shall be used to identify the image data contained in this record. The IDC contained in this field shall be a binary representation of the IDC found in the file content (CNT) field of the Type-1 record.

11.1.3 Impression type (IMP)

This mandatory one-byte binary field shall occupy the sixth byte of a Type-4 record. The code selected from Table 5, describing the manner by which the fingerprint image information was obtained, shall be entered in this field.

11.1.4 Finger position (FGP)

This mandatory fixed-length field of 6 binary bytes shall occupy the seventh through twelfth positions of a Type-4 record. It shall contain possible finger positions beginning in the leftmost byte of the field (byte seven of the record). The decimal code number for the known or most probable finger position shall be taken from Table 6 and entered as a binary number right justified and left zero filled within the eight-bit byte. Up to five additional finger positions may be referenced by entering the alternate finger positions in the remaining five bytes using the same format. If fewer than five finger position references are to be used, the unused bytes shall be filled with the binary equivalent of "255". The code "0", for "Unknown Finger", shall be used to reference every finger position from one through ten.

11.1.5 Image scanning resolution (ISR)

This mandatory one-byte binary field shall occupy the thirteenth byte of a Type-4 record. It shall contain a binary value of "0" if the minimum scanning resolution is used, and a "1" if the native scanning resolution is used.

11.1.6 Horizontal line length (HLL)

This mandatory two-byte binary field shall occupy the fourteenth and fifteenth bytes of the Type-4 record. It shall be used to specify the number of pixels contained on a single horizontal line of the transmitted image.

11.1.7 Vertical line length (VLL)

This mandatory two-byte binary field shall occupy the sixteenth and seventeenth bytes of the Type-4 record. It shall be used to specify the number of horizontal lines contained in the transmitted image.

11.1.8 Grayscale compression algorithm (GCA)

This mandatory one-byte binary field shall occupy the eighteenth byte of a Type-4 record. It shall be used to specify the type of grayscale compression algorithm used (if any). A binary zero denotes no compression. Otherwise, the contents of this byte shall be a binary representation for the number allocated to the particular compression technique used by the interchange parties. The domain registrar will maintain a registry relating these numbers to the compression algorithms.

11.1.9 Image data

This binary field shall contain all of the high-resolution grayscale image data. Each pixel of the uncompressed image shall be quantized to eight bits (256 gray levels) contained in a single byte. If compression is used, the pixel data shall be compressed in accordance with the compression technique specified in the GCA field. This completes the high-resolution image description for a single image.

11.2 End of Type-4 high-resolution grayscale fingerprint image record

Since the Type-4 logical record is a defined and specified series of binary data fields, no additional bytes shall be transmitted to signify the end of this logical record type.

11.3 Additional high-resolution grayscale fingerprint images

Typically, up to thirteen more images may be described within the file. For each additional image, a Type-4 logical record is required.

12 Type-5 low-resolution binary fingerprint image record

Type-5 logical records shall contain low-resolution binary fingerprint image data. The fingerprint image data shall have been scanned at no less than the minimum scanning resolution and then subsampled, scaled down, or interpolated. Alternatively, provided that it is no less than the minimum scanning resolution, the native scanning resolution may be used and the image processed such that the resulting transmitting resolution is within the range specified by the transmitting resolution requirement for low-resolution images. When the image data are obtained from a live-scan reader, it shall be the binarized subsampled, scaled down, or interpolated output of the live-scan fingerprint scanner and not a rescan of a hard copy fingerprint image.

12.1 Fields for Type-5 logical record

Within a Type-5 logical record, entries shall be provided in nine ordered and unnumbered fields. The first eight fields are fixed length and total eighteen bytes. These fields precede the image data contained in field nine. The size of the ninth field is eighteen bytes less than the value specified in the LEN field.

12.1.1 Logical record length (LEN)

This mandatory four-byte binary field shall occupy bytes one through four. It shall contain the length of the logical record specifying the total number of bytes, including every byte of all nine fields contained in the record.

12.1.2 Image designation character (IDC)

This mandatory one-byte binary field shall occupy the fifth byte of a Type-5 record. It shall be used to identify the image data contained in this record. The IDC contained in this field shall be a binary representation of the IDC found in the file content (CNT) field of the Type-1 record.

12.1.3 Impression type (IMP)

This mandatory one-byte binary field shall occupy the sixth byte of a Type-5 record. The code selected from Table 5, describing the manner by which the fingerprint image information was obtained, shall be entered in this field.

12.1.4 Finger position (FGP)

This mandatory fixed-length field of six binary bytes shall occupy the seventh through twelfth positions of a Type-5 record. It shall contain possible finger positions beginning in the leftmost byte of the field (byte seven of the record). The decimal code number for the known or most probable finger position shall be taken from Table 6 and entered as a binary number right justified and left zero filled within the eight-bit byte. Up to five additional finger positions may be referenced by entering the alternate finger positions in the remaining five bytes using the same format. If fewer than five finger position references are to be used, the unused bytes shall be filled with the binary equivalent of "255". The code "0", for "Unknown Finger", shall be used to reference every finger position from one through ten.

12.1.5 Image scanning resolution (ISR)

This mandatory one-byte binary field shall occupy the thirteenth byte of a Type-5 record. It shall contain a binary value of "0" if half the minimum scanning resolution is used, and a "1" if half the native scanning resolution is used.

12.1.6 Horizontal line length (HLL)

This mandatory two-byte binary field shall occupy the fourteenth and fifteen bytes of the Type-5 record. It shall be used to specify the number of pixels contained on a single horizontal line of the transmitted image.

12.1.7 Vertical line length (VLL)

This mandatory two-byte binary field shall occupy the sixteenth and seventeenth bytes of the Type-5 record. It shall be used to specify the number of horizontal lines contained in the transmitted image.

12.1.8 Binary compression algorithm (BCA)

This mandatory one-byte binary field shall occupy the eighteenth byte of a Type-5 record. It shall be used to specify whether or not data compression is used. A binary zero denotes no compression. A binary one denotes the use of the ANSI/EIA-538-1988 facsimile compression standard for the lossless compression and decompression of the image data.

12.1.9 Image data

This field shall contain all of the low-resolution binary image data. Each pixel of the uncompressed image shall be quantized to two levels (binary representation). For binary pixels, a value of "0" shall represent a white pixel and a value of "1" shall represent a black pixel. Uncompressed data shall be packed at eight pixels per byte. If compression is used, the pixel data shall be compressed in accordance with the technique as specified in BCA field. This completes the low-resolution binary image description for a single image.

12.2 End of Type-5 low-resolution binary fingerprint image record

Since the Type-5 logical record is a defined and specified series of binary data fields, no additional bytes shall be transmitted to signify the end of this logical record type.

12.3 Additional low-resolution binary fingerprint image records

Typically, up to thirteen more images may be described within the file. For each additional image, a Type-5 logical record is required.

13 Type-6 high-resolution binary fingerprint image record

Type-6 logical records shall contain high-resolution binary fingerprint image data that has been scanned at the minimum scanning resolution. Alternatively, the native scanning resolution may be used and the image processed such that the resulting transmitting resolution is within the range specified by the transmitting resolution requirement for high-resolution images. When the image data are obtained from a live-scan reader it shall be the binarized output of the live-scan fingerprint scanner and not a rescan of a hard copy fingerprint image.

13.1 Fields for Type-6 logical record

Within a Type-6 logical record, entries shall be provided in nine ordered and unnumbered fields. The first eight fields are fixed length and total eighteen bytes. These fields precede the image data contained in field nine. The size of the ninth field is eighteen bytes less than the value specified in the LEN field.

13.1.1 Logical record length (LEN)

This mandatory four-byte binary field shall occupy bytes one through four. It shall contain the length of the logical record specifying the total number of bytes, including every byte of all nine fields contained in the record.

13.1.2 Image designation character (IDC)

This mandatory one-byte binary field shall occupy the fifth byte of a Type-6 record. It shall be used to identify the image data contained in this record. The IDC contained in this field shall be a binary representation of the IDC found in the file content (CNT) field of the Type-1 record.

13.1.3 Impression type (IMP)

This mandatory one-byte binary field shall occupy the sixth byte of a Type-6 record. The code selected from Table 5, describing the manner by which the fingerprint image information was obtained, shall be entered in this field.

13.1.4 Finger position (FGP)

This mandatory fixed-length field of six binary bytes shall occupy the seventh through twelfth positions of a Type-6 record. It shall contain possible finger positions beginning in the leftmost byte of the field (byte seven of the record). The decimal code number for the known or most probable finger position shall be taken from Table 6 and entered as a binary number right justified and left zero filled within the eight-bit byte. Up to five additional finger positions may be referenced by entering the alternate finger positions in the remaining five bytes using the same format. If fewer than five finger position references are to be used, the unused bytes shall be filled with the binary equivalent of "255". The code "0", for "Unknown Finger", shall be used to reference every finger position from one through ten.

13.1.5 Image scanning resolution (ISR)

This mandatory one-byte binary field shall occupy the thirteenth byte of a Type-6 record. It shall contain a binary value of "0" if the minimum scanning resolution is used and a "1" if the native scanning resolution is used.

13.1.6 Horizontal line length (HLL)

This mandatory two-byte binary field shall occupy the fourteenth and fifteenth bytes of the Type-6 record. It shall be used to specify the number of pixels contained on a single horizontal line of the transmitted image.

13.1.7 Vertical line length (VLL)

This mandatory two-byte binary field shall occupy the sixteenth and seventeenth bytes of the Type-6 record. It shall be used to specify the number of horizontal lines contained in the transmitted image.

13.1.8 Binary compression algorithm (BCA)

This mandatory one-byte binary field shall occupy the eighteenth byte of a Type-6 record. It shall be used to specify whether or not data compression is used. A binary zero denotes no compression. A binary one denotes the use of the ANSI/EIA-538-1988 facsimile compression standard for the lossless compression and decompression of the image data.

13.1.9 Image data

This field shall contain all of the high-resolution binary image data. Each pixel of the uncompressed image shall be quantized to two levels (binary representation). For binary pixels, a value of "0" shall represent a white pixel and a value of "1" shall represent a black pixel. Uncompressed data shall be packed at eight pixels per byte. If compression is used, pixel data shall be compressed in accordance with the technique as specified in BCA field. This completes the high-resolution binary image description for a single image.

13.2 End of Type-6 high-resolution binary fingerprint image record

Since the Type-6 logical record is a defined and specified series of binary data fields, no additional bytes shall be transmitted to signify the end of this logical record type.

13.3 Additional high-resolution binary fingerprint image records

Typically, up to thirteen more images may be described within the file. For each additional image, a Type-6 logical record is required.

14 Type-7 user-defined image record

Type-7 logical records shall contain user-defined image information relating to the transaction submitted for processing. This record type is designed to be used to exchange image data that is not addressed elsewhere in this standard. Examples of usage include, but are not limited to, images pertaining to the toes and soles of a subject's feet. Scanned pixels may be either binary or grayscale output. Each grayscale pixel value shall be expressed as an unsigned byte. A value of "0" shall be used to define a black pixel and an unsigned value of "255" shall be used to define a white pixel. For binary pixels, a value of "0" shall represent a white pixel and a value of "1" shall represent a black pixel. If compression is used, the algorithm shall be the same as that specified for Type-3, Type-4, Type-5, and Type-6 logical records.

14.1 Fields for Type-7 logical record

The Type-7 logical record is a binary record that shall not contain any ASCII data. The first two data fields of the Type-7 record are defined by this Standard. Remaining fields of the record shall conform to the requirements set forth by the agency receiving the transmission.

The first two fields are fixed length and total five bytes. These fields shall precede one or more user-defined fields, including the image data, contained in the remainder of the record.

14.1.1 Logical record length (LEN)

This mandatory four-byte binary field shall occupy bytes one through four. It shall contain the length of the logical record specifying the total number of bytes, including every byte of all the fields contained in the record.

14.1.2 Image designation character (IDC)

This mandatory one-byte binary field shall occupy the fifth byte of a Type-7 record. It shall be used to identify the image data contained in this record. The IDC contained in this field shall be a binary representation of the IDC found in the file content (CNT) field of the Type-1 record.

14.1.3 User-defined fields for Type-7 logical record

The remaining fields of the Type-7 logical record shall be user-defined. Individual fields required for a given transaction, such as field description, size, and content shall conform to the specifications set forth by the agency to whom the transmission is being sent.

14.2 End of Type-7 user-defined image record

Since the Type-7 logical record is a defined and specified series of binary data fields, no additional bytes shall be transmitted to signify the end of this logical record type.

14.3 Additional user-defined image records

Additional images may be described within the file. For each additional image, a Type-7 logical record is required.

15 Type-8 signature image record

Type-8 logical records shall contain either scanned or vectored signature data. Each Type-8 record shall cover an area of up to 1000 mm^2.

If scanned, the resolution shall be the minimum scanning resolution or the native scanning resolution, and the scan sequence shall be left to right and top to bottom. The scanned data shall be a binary representation quantized to two levels.

If vectored signature data is present, it shall be expressed as a series of binary numbers.

15.1 Fields for Type-8 logical record

When there are one or two Type-8 logical records, entries shall be provided in eight ordered and unnumbered binary fields. The first seven fields are fixed length and shall total twelve bytes. These fields shall precede the image data contained in field eight. The size of field eight is determined from the LEN field of the record itself. The image data field is 12 bytes less than the value specified in the LEN field.

15.1.1 Logical record length (LEN)

This mandatory four-byte binary field shall occupy bytes one through four. It shall contain the length of the logical record expressed as the total number of bytes, including every byte of all eight fields contained in the record.

15.1.2 Image designation character (IDC)

This mandatory one-byte binary field shall occupy the fifth byte of the Type-8 record. It shall be used to identify the image data contained in the Type-8 record. The IDC contained in this field shall be a binary representation of the IDC found in the file content (CNT) field of the Type-1 record.

15.1.3 Signature type (SIG)

This mandatory one-byte binary field shall occupy the sixth byte of the Type-8 record. It shall contain a binary "0" for the signature image of the subject, or a binary "1" for the signature image of the official processing the transaction.

15.1.4 Signature representation type (SRT)

This mandatory one-byte binary field shall occupy the seventh byte of the Type-8 record. Its value shall be a binary "0" if the image is scanned and not compressed, a binary "1" if the image is scanned and compressed, and the binary equivalent of "2" if the image is vector data.

15.1.5 Image scanning resolution (ISR)

This mandatory one-byte binary field shall occupy the eighth byte of a Type-8 record. It shall contain a binary "0" if the minimum scanning resolution is used and a binary "1" if the native scanning resolution is used. A binary value of "0" shall also be used if the image is vector data.

15.1.6 Horizontal line length (HLL)

This mandatory two-byte binary field shall occupy the ninth and tenth bytes of the Type-8 record. For scanned signature data, this field shall be used to specify the number of pixels contained on a single horizontal line of the transmitted signature image. For vectored signature data, both bytes shall contain the binary value of "0".

15.1.7 Vertical line length (VLL)

This mandatory two-byte binary field shall occupy the eleventh and twelfth bytes of the Type-8 record. For scanned signature data, this field shall be used to specify the number of horizontal lines contained in the transmitted signature image. For vectored signature data, both bytes shall contain the binary value of "0".

15.1.8 Signature image data

This field shall contain uncompressed scanned image signature data, compressed scanned image signature data, or vectored image signature data. The entry contained in the SRT field shall indicate which form of the signature data is present.

15.1.8.1 Uncompressed scanned image data

If the SRT field contains the binary value of "0", then this field shall contain the uncompressed scanned binary image data for the signature. In uncompressed mode, the data shall be packed at eight pixels per byte.

15.1.8.2 Compressed scanned image data

If the SRT field contains the binary value of "1", then this field shall contain the scanned binary image data for the signature in compressed form using the ANSI/EIA-538-1988 facsimile compression algorithm.

15.1.8.3 Vectored image data

If the SRT field contains the binary equivalent of "2", then this field shall contain a list of vectors describing the pen position and pen pressure of line segments within the signature. Each vector shall consist of five bytes.

The first two bytes of each vector shall contain the unsigned binary X coordinate of the pen position with the high order byte containing the most significant bits. The next two bytes shall contain the unsigned Y coordinate using the same convention to denote the most significant bits. Both the X and Y coordinates shall be expressed in units of .0254 mm (.001 inches) referenced from the bottom leftmost corner of the signature. Positive values of X shall increase from left-to-right and positive values of Y shall increase from bottom-to-top.

An unsigned binary number between "0" and "255" contained in the fifth byte shall represent the pen pressure. This shall be a constant pressure until the next vector becomes active. A binary value or pressure of "0" shall represent a "pen-up" (or no pressure) condition. The binary value of "1" shall represent the least recordable pressure for a particular device, while the binary equivalent of "254" shall represent the maximum recordable pressure for that device. To denote the end of the vector list the binary equivalent of "255" shall be inserted in this entry.

15.2 End of Type-8 signature image record

Since the Type-8 logical record is a defined and specified series of binary data fields, no additional bytes shall be transmitted to signify the end of this logical record type.

15.3 Additional signature

One more signature may be described within the file. For an additional signature, a Type-8 logical record is required.

16 Type-9 minutiae data record

Type-9 records shall contain ASCII text describing minutiae and related information encoded from a finger or palm. For a tenprint search transaction, there may be up to ten of these Type-9 records in a file, each of which shall be for a different finger. There may be up to six of these records for palmprint searches. The Type-9 record shall also be used to exchange the minutiae information from latent finger or palm images between similar or different systems.

16.1 Minutiae and other information descriptors

16.1.1 Minutia type identification

This standard defines four identifier characters that are used to describe the minutia type. These are listed in Table 7. A ridge ending shall be designated Type A. It occurs at the point on a fingerprint or palmprint that a friction ridge begins or ends without splitting into two or more continuing ridges. The ridge must be longer than it is wide. A bifurcation shall be designated Type B. It occurs at the point that a ridge divides or splits to form two ridges that continue past the point of division for a distance that is at least equal to the spacing between adjacent ridges at the point of bifurcation. A minutia shall be designated Type C, a compound type, if it is either a trifurcation (a single ridge that splits into three ridges) or a crossover (two ridges that intersect). If a minutia cannot be clearly categorized as one of the above three types, it shall be designated as undetermined, Type D.

Table 7 – Minutia types

Type	Description
A	Ridge ending
B	Bifurcation
C	Compound (trifurcation or crossover)
D	Type undetermined

16.1.2 Minutia numbering

Each minutia shall be identified by an index number that is assigned to it. The numbering shall begin at "1" and be incremented by "1" for as many times as there are minutiae encountered. This allows each minutia to be uniquely identified.

16.1.3 Minutiae ridge counts

As required, ridge counts shall be determined from each minutia in a fingerprint or palmprint to certain other neighboring minutiae. When this occurs, ridge counts between designated minutiae shall be associated with the applicable index numbers so as to ensure maintenance of the proper relationships. Rules for identifying neighboring minutiae and the method to be used for counting the intervening ridge crossings is not part of this standard.

16.1.4 Minutiae coordinate system

The relative position of minutiae entered in Type-9 records shall be expressed as positive integers in units of 0.01 mm (0.00039 in) in a Cartesian coordinate system located in Quadrant 1. In this coordinate system, values of X increase from left to right and values of Y increase from bottom to top.

For encoded minutiae from fingerprints, values of both X and Y are equal to or greater than "0000" and are less than "5000". This range of units converts to 5 cm (1.97") in both the horizontal and vertical directions. If the conversion to this coordinate system is from a system that normally centers the fingerprint image during the registration process, that center position shall be assigned the values X = 2500, Y = 2500. Figure 2 illustrates the defined coordinate system for a fingerprint.

For encoded minutiae from a palmprint, values of both X and Y are equal to or greater than "0000" and are less than "14000" and "21000" respectively. This range of units converts to 14 cm (5.51") in the horizontal and 21 cm (8.27") in the vertical directions.

The relative orientation, Theta, of a ridge ending, a bifurcation, a compound or a minutia of undetermined type shall be expressed as positive integers in units of degrees from "0" to "359" degrees. Theta shall be the angle between the horizontal axis of the coordinate system and the direction that a ridge ending points, assuming that a ridge ending is analogous to a pointing finger. A ridge ending that is formed by a ridge lying parallel to the X axis, and ending in the direction of increasing values of X, shall have an orientation of zero degrees. Counterclockwise rotation of this ridge about the ridge ending shall cause the value of Theta to increase. A ridge ending pointing due east has a direction of zero degrees, due north 90 degrees and so forth.

A bifurcation may be converted to a ridge ending by logical inversion, i.e., transposing the identity of ridges and valleys. The orientation of a bifurcation is expressed as if this inversion had occurred. This convention causes no significant change in the orientation of a minutia if it appears as a ridge ending in one impression of a fingerprint and as a bifurcation in another impression of the same fingerprint.

No orientation shall be assigned to a compound type minutiae; therefore, a value of "000" shall be entered for Theta in the Type-9 logical record entry.

The exact features or characteristics of a minutia that are used to establish its position and orientation are system dependent and outside the scope of this standard.

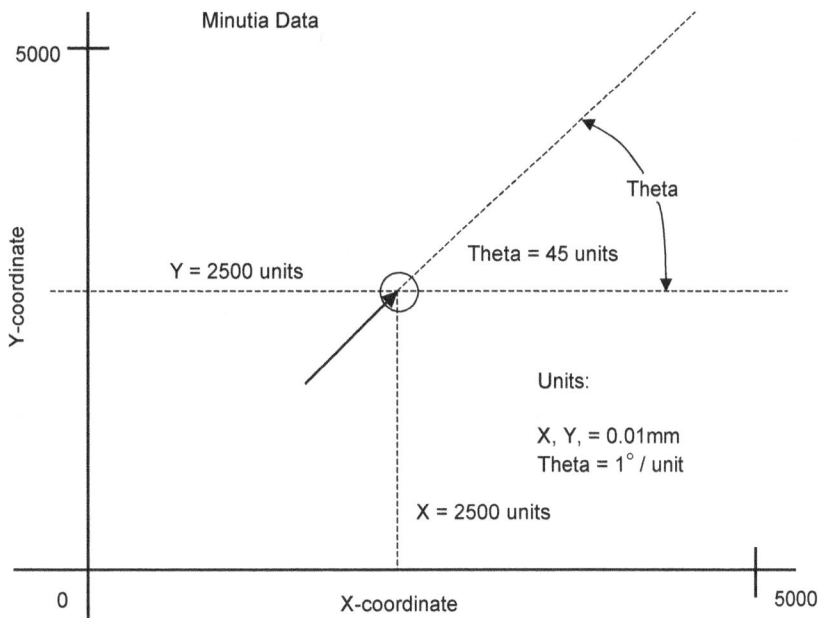

Figure 2 - Minutiae coordinate system

16.2 Fields for Type-9 logical record

All fields of the Type-9 records shall be recorded as ASCII text. No binary fields are permissible in this tagged-field record. The first twelve ASCII fields of the Type-9 logical record provide a common or generic manner of encoding minutiae and other characteristic data. These fields are formatted in accordance with the conventions described above.

This logical record type can also be used to accommodate a variety of methods used by AFIS vendors for encoding minutiae data according to their particular requirements. The numbered fields of the Type-9 logical record are partitioned into registered blocks of tagged-fields starting at Field-13. Each AFIS vendor has a block of uniquely numbered fields reserved for the encoding of minutiae and other characteristic data required for their feature vector. Each specific vendor implementation must contain the first four fields described below. The remaining numbered fields of the feature vector are vendor dependent and encoded according to their own conventions. None of the fields from Field-5 through Field-12 described below are required to be present in specific vendor implementations.

16.2.1 Field 9.001: Logical record length (LEN)

This mandatory ASCII field shall contain the length of the logical record specifying the total number of bytes, including every character of every field contained in the record.

16.2.2 Field 9.002: Image designation character (IDC)

This mandatory field shall be used for the identification and location of the minutiae data. The IDC contained in this field shall match the IDC found in the file content (CNT) field of the Type-1 record.

16.2.3 Field 9.003: Impression type (IMP)

This mandatory one-byte field shall describe the manner by which the fingerprint image information was obtained. The ASCII value of the proper code as selected from Table 5 or Table 18 shall be entered in this field to signify the impression type.

16.2.4 Field 9.004: Minutiae format (FMT)

This mandatory one-byte field shall be used to indicate whether the information in the remainder of the record adheres to the standard format or is a user-defined format. This field shall contain an "S" to indicate that the minutiae are formatted as specified by the standard Type-9 logical record field descriptions using location information and other conventions described above. A standard Type-9 logical record will use Field-5 through Field-12 as described below.

This field shall contain a "U" to indicate that the minutiae are formatted in vendor-specific terms. This alternative allows each vendor to encode minutiae data and any additional required characteristic or feature data in accordance with their own system's specific hardware and software configuration. Reserved blocks, each consisting of several tagged fields, are registered to and allocated for use by specific vendors. Vendor-specific data shall begin with Field-13. The assignment of blocks of tagged-fields to specific vendors is controlled by the domain registrar responsible for the implementation domain. By default this shall be the registrar for the North American Domain. Multiple blocks of vendor-specific data, which may include Field-5 through Field-12, may occur within a single Type-9 record when this field contains a "U".

Even though information may be encoded in accordance with a specific vendor's implementation, all data fields of the Type-9 record must remain as ASCII text fields.

16.2.5 Field 9.005: Originating fingerprint reading system (OFR)

The originator's designation or name for the particular fingerprint or palmprint reading system that generated this record shall be placed in the first information item of this field. The second information item of this field shall be a single character to indicate the method by which the minutiae data was read, encoded, and recorded. The following coding shall be used: (1) "A", if the data was automatically read, encoded, and recorded without any possibility of human editing; (2) "U", if human editing was possible but unneeded; (3) "E", if the data was automatically read but manually edited before encoding and recording; (4) "M", if the data was manually read. The third information item is an optional, two-character, user-generated subsystem designator that uniquely identifies the originator's equipment.

16.2.6 Field 9.006: Finger position (FGP)

This mandatory field shall contain the code designating the finger or palm position that produced information in this Type-9 record. If the exact finger or palm position cannot be determined, multiple finger positions may be entered, separated by the *"RS"* character. Entries from Table 6 or Table 19 list the codes that shall be used for each fingerprint or palmprint.

16.2.7 Field 9.007: Fingerprint pattern classification (FPC)

This mandatory field shall contain the fingerprint pattern classification code. It shall contain two information items. The first information item shall indicate the source of the specific pattern classification code. It may be one chosen from Table 8 or may be a user-defined classification code. This item shall contain a "T" to indicate that the pattern classification code is from Table 8, or a "U" to indicate that the code is user-defined. The second information item of this field shall contain the pattern classification code chosen from Table 8 or a specific user-defined code. When it is not possible to uniquely identify the fingerprint class, reference fingerprint classes may be used and shall be separated by the *"RS"* character.

Table 8 – Pattern classification

Description	Code
Plain arch	PA
Tented arch	TA
Radial loop	RL
Ulnar loop	UL
Plain whorl	PW
Central pocket loop	CP
Double loop	DL
Accidental whorl	AW
Whorl, type not designated	WN
Right slant loop	RS
Left slant loop	LS
Scar	SR
Amputation	XX
Unknown or unclassifiable	UN

16.2.8 Field 9.008: Core position (CRP)

If this eight-character field is used, it shall contain the X and Y coordinate position of the core of a fingerprint. The X and Y values shall be coded as a single 8-digit integer number comprised of the 4-digit X-coordinate concatenated with the 4-digit Y-coordinate using a format of XXXXYYYY. The *"RS"* separator shall separate multiple occurrences of core positions

16.2.9 Field 9.009: Delta(s) position (DLT)

If this eight-character field is used, it shall contain the X and Y positional coordinates of each delta that is present on the fingerprint. The X and Y values shall be recorded in the same manner as was done for the core position coordinates. The *"RS"* separator shall separate multiple occurrences of delta positions.

16.2.10 Field 9.010: Number of minutiae (MIN)

This mandatory textual field shall contain the count of the number of minutiae recorded for this fingerprint or palmprint.

16.2.11 Field 9.011: Minutiae ridge count indicator (RDG)

This mandatory single-character field shall be used to indicate the presence of minutiae ridge count information. A "0" in this field indicates that no ridge count information is available. A "1" indicates that ridge count information is available.

16.2.12 Field 9.012: Minutiae and ridge count data (MRC)

This variable field length shall contain all of the individual minutiae and ridge count data associated with the current fingerprint impression. It shall be comprised of as many subfields as there are minutiae stated in the minutiae count in Field 9.010. Each subfield shall be devoted to a single minutia and shall consist of multiple information items. The first two information items shall always appear; the appearance of others is system dependent. The information items are identified in the order that they shall appear. All information items shall be separated from the subsequent items by the *"US"* separator character.

16.2.12.1 Index number

The first information item shall be the index number, which shall be initialized to "1" and incremented by "1" for each additional minutia in the fingerprint. This index number serves to identify each individual minutia.

16.2.12.2 X, Y, and theta values

For minutiae encoded from fingerprints, the X and Y coordinates (two 4-digit values ranging from zero upward), and the Theta value (a 3-digit value between 000 and 359) shall comprise the second required information item. These three values shall be coded and recorded as a single 11-digit integer number corresponding to the concatenated X, Y, and Theta values, in that order.

For minutiae encoded from palmprints, the X and Y coordinates (two 5-digit values ranging from zero upward), and the three-digit Theta value shall comprise the second required information item. These three values shall be coded and recorded as a single 13-digit integer number corresponding to the concatenated X, Y, and Theta values, in that order.

16.2.12.3 Quality measure

If present, the third information item is an optional quality measure. Values shall range from "0" to "63". The value "0" shall indicate a manually encoded minutia. The value "1" shall indicate that no method of indicating a confidence level is available. Values between "2" and "63" shall indicate decreasing levels of confidence, with "2" meaning the greatest confidence. If the quality measure information item is not available for this minutia but the type and/or ridge count data is present, then a "US" information separator character must be included.

16.2.12.4 Minutia type designation

The fourth information item is an optional minutia type designation. This shall be a single alphabetic character as chosen from Table 7. If the minutia type information is not available for this minutia but ridge count data is present, then a "US" information separator character must be included.

16.2.12.5 Ridge count data

The fifth information item is optional ridge count data. It shall be formatted as a series of information items, each consisting of a minutia number and a ridge count. This information shall be conveyed by listing the identity (index number) of the distant minutia followed by a comma, and the ridge count to that distant minutia. The *"US"* character shall be used to separate these information items. These information items shall be repeated as many times as required for each minutia (subfield).

16.2.12.6 Record separator

A Record Separator character, *"RS"*, shall be used at the end of the information items to introduce the first information item concerning data for the next minutia. The process shall be continued until all of the minutiae and ridge data have been entered into the field.

16.3 End of Type-9 logical record

Immediately following the last information field in the Type-9 logical record, an *"FS"* separator shall be used to separate it from the next logical record or signify the end of the transaction. This separator character must be included in the length field of the Type-9 record.

16.4 Additional minutiae records

Typically, up to nine more fingers may be described within the file. For each additional finger, a Type-9 logical record, and an *"FS"* separator is required.

17 Type-10 facial & SMT image record

Type-10 records shall contain facial and/or SMT image data and related ASCII information pertaining to the specific image contained in this record. It shall be used to exchange both grayscale and color image data either in a compressed or an uncompressed form.

17.1 Fields for Type-10 logical record

The following paragraphs describe the data contained in each of the fields for the Type-10 logical record.

Within a Type-10 logical record, entries shall be provided in numbered fields. It is required that the first two fields of the record are ordered, and the field containing the image data shall be the last physical field in the record. For each field of the Type-10 record, Table 9 lists the "condition code" as being mandatory "M" or optional "O", the field number, the field name, character type, field size, and occurrence limits. Based on a three digit field number, the maximum byte count size for the field is given in the last column. As more digits are used for the field number, the maximum byte count will also increase. The two entries in the "field size per occurrence" include all character separators used in the field. The "maximum byte count" includes the field number, the information, and all the character separators. Fields containing entries in the "IMG" column are only applicable to that image type. An entry of "FAC" applies to a mugshot or facial image, and an entry of "SMT" applies to scar, a mark, or a tattoo image.

17.1.1 Field 10.001: Logical record length (LEN)

This mandatory ASCII field shall contain the total count of the number of bytes in the Type-10 logical record. Field 10.001 shall specify the length of the record including every character of every field contained in the record and the information separators.

17.1.2 Field 10.002: Image designation character (IDC)

This mandatory ASCII field shall be used to identify the facial or SMT image data contained in the record. This IDC shall match the IDC found in the file content (CNT) field of the Type-1 record.

17.1.3 Field 10.003: Image type (IMT)

This mandatory ASCII field is used to indicate the type of image contained in this record. It shall contain *"FACE"*, *"SCAR"*, *"MARK"*, or *"TATTOO"* to indicate the appropriate image type.

ANSI/NIST-ITL 1-2000

Table 9 – Type-10 facial and SMT record layout

Ident	Cond code	Field Number	Field Name	IMG	Char type	Field size per occurrence		Occur count		Max byte count
						min	max	min	max	
LEN	M	10.001	LOGICAL RECORD LENGTH		N	4	8	1	1	15
IDC	M	10.002	IMAGE DESIGNATION CHARACTER		N	2	5	1	1	12
IMT	M	10.003	IMAGE TYPE		A	5	7	1	1	14
SRC	M	10.004	SOURCE AGENCY / ORI		AN	10	21	1	1	28
PHD	M	10.005	PHOTO DATE		N	9	9	1	1	16
HLL	M	10.006	HORIZONTAL LINE LENGTH		N	4	5	1	1	12
VLL	M	10.007	VERTICAL LINE LENGTH		N	4	5	1	1	12
SLC	M	10.008	SCALE UNITS		N	2	2	1	1	9
HPS	M	10.009	HORIZONTAL PIXEL SCALE		N	2	5	1	1	12
VPS	M	10.010	VERTICAL PIXEL SCALE		N	2	5	1	1	12
CGA	M	10.011	COMPRESSION ALGORITHM		A	5	7	1	1	14
CSP	M	10.012	COLOR SPACE		A	4	5	1	1	12
RSV	-	10.013 10.019	RESERVED FOR FUTURE INCLUSION		--	--	--	--	--	--
POS	O	10.020	SUBJECT POSE	FAC	A	2	2	0	1	9
POA	O	10.021	POSE OFFSET ANGLE	FAC	N	2	5	0	1	12
PXS	O	10.022	PHOTO DESCRIPTION	FAC	A	4	21	0	9	196
RSV	-	10.023 10.039	RESERVED FOR FUTURE INCLUSION		--	--	--	--	--	---
SMT	M	10.040	NCIC DESIGNATION CODE	SMT	A	4	11	1	3	40
SMS	O	10.041	SCAR/MARK/TATTOO SIZE	SMT	N	4	6	0	1	13
SMD	O	10.042	SMT DESCRIPTORS	SMT	AN	16	51	0	9	466
COL	O	10.043	COLORS PRESENT	SMT	A	4	21	0	9	196
RSV	-	10.044 10.199	RESERVED FOR FUTURE INCLUSION		--	--	--	--	--	---
UDF	O	10.200 10.998	USER-DEFINED FIELDS		--	--	--	--	--	---
DAT	M	10.999	IMAGE DATA		B	2	5,000,001	1	1	5,000,008

Key for character type: N = Numeric; A = Alphabetic; AN = Alphanumeric; B = Binary

17.1.4 Field 10.004: Source agency / ORI (SRC)

This mandatory ASCII field shall contain the identification of the administration or organization that originally captured the facial image contained in the record. Normally, the Originating Agency Identifier, ORI, of the agency that captured the image will be contained in this field. The user initiating the transaction shall define the size of this field and its data content so that it can be processed by the receiving agency.

17.1.5 Field 10.005: Photo date (PHD)

This mandatory ASCII field shall contain the date that the facial or SMT image contained in the record was captured. The date shall appear as eight digits in the format *CCYYMMDD*. The *CCYY* characters shall represent the year the image was captured; the *MM* characters shall be the tens and units values of the month; and the *DD* characters shall be the tens and units values of the day in the month. For example, 20000229 represents February 29, 2000. The complete date must be a legitimate date.

17.1.6 Field 10.006: Horizontal line length (HLL)

This mandatory ASCII field shall contain the number of pixels contained on a single horizontal line of the transmitted image.

17.1.7 Field 10.007: Vertical line length (VLL)

This mandatory ASCII field shall contain the number of horizontal lines contained in the transmitted image.

17.1.8 Field 10.008: Scale units (SLC)

This mandatory ASCII field shall specify the units used to describe the image sampling frequency (pixel density). A "1" in this field indicates pixels per inch, or a "2" indicates pixels per centimeter. A "0" in this field indicates no scale is given. For this case, the quotient of HPS/VPS gives the pixel aspect ratio.

17.1.9 Field 10.009: Horizontal pixel scale (HPS)

This mandatory ASCII field shall specify the integer pixel density used in the horizontal direction providing the SLC contains a "1" or a "2". Otherwise, it indicates the horizontal component of the pixel aspect ratio.

17.1.10 Field 10.010: Vertical pixel scale (VPS)

This mandatory ASCII field shall specify the integer pixel density used in the vertical direction providing the SLC contains a "1" or a "2". Otherwise, it indicates the vertical component of the pixel aspect ratio.

17.1.11 Field 10.011: Compression algorithm (CGA)

This mandatory ASCII field shall specify the algorithm used to compress the color or grayscale image. An entry of *"NONE"* in this field indicates that the data contained in this record is uncompressed. For those images that are to be compressed, the preferred method for the compression of facial and SMT images is specified by the baseline mode of the JPEG algorithm. The data shall be formatted in accordance with the JPEG File Interchange Format, Version 1.02 (JFIF)[7] as found in Annex D. An entry of *"JPEGB"* indicates that the scanned or captured image was compressed using baseline JPEG. An entry of *"JPEGL"* indicates that the lossless mode of the JPEG algorithm was used to compress the image. If the image is captured in grayscale, then only the luminescence component will be compressed and transmitted.

17.1.12 Field 10.012: Colorspace (CSP)

This mandatory ASCII field shall contain the color space used to exchange the image. For compressed images, the preferred colorspace using baseline JPEG and JFIF is YCbCr[8] to be coded as *"YCC"*. An entry of *"GRAY"* shall be used for all grayscale images. This field shall contain *"RGB"* for uncompressed color images containing non-interleaved red, green, and blue pixels in that order. All other colorspaces are undefined.

17.1.13 Field 10.013-019: Reserved for future definition (RSV)

These fields are reserved for inclusion in future revisions of this standard. None of these fields are to be used at this revision level. If any of these fields are present, they are to be ignored.

17.1.14 Field 10.020: Subject pose (POS)

This optional field is to be used for the exchange of facial image data. When included, this field shall contain one ASCII character code selected from Table 10 to describe the pose of the subject. For the angled pose entry "A", field 10.021 shall contain the offset angle from the full face orientation.

Table 10 – Subject pose

Pose description	Pose code
Full Face Frontal	F
Right Profile (90 degree)	R
Left Profile (90 degree)	L
Angled Pose	A

---

[7] Developed by C-Cube Microsystems, 1778 McCarthy Blvd., Milpitas, CA 95035.

[8] Annex D contains the information necessary to perform conversions between 24-bit RGB pixels and the YCbCr color space.

17.1.15 Field 10.021: Pose offset angle (POA)

This field shall only be used for the exchange of facial image data if Field 10.020 (POS) contains an *"A"* to indicate an angled pose of the subject. This field should be omitted for a full face or a profile. This ASCII field specifies the pose position of the subject at any possible orientation within a circle. Its value shall be to a nearest degree.

The offset angle shall be measured from the full-face pose position and have a range of values from -180 degrees to +180 degrees. A positive angle is used to express the angular offset as the subject rotates from a full-face pose to their right (approaching a left profile). A negative angle is used to express the angular offset as the subject rotates from a full-face pose to their left (approaching a right profile). If the entry in the POS field is an *"F"*, *"L"*, or *"R"*, the contents of this field are ignored.

17.1.16 Field 10.022: Photo description (PXS)

This optional ASCII field shall be used for the exchange of facial image data. When present, it shall describe special attributes of the captured facial image. Attributes associated with the facial image may be selected from Table 11 and entered in this field.

Table 11 – Photo descriptors

Facial image attribute	Attribute code
Subject Wearing Glasses	GLASSES
Subject Wearing Hat	HAT
Subject Wearing Scarf	SCARF
Physical Characteristics	PHYSICAL
Other Characteristics	OTHER

Physical characteristics, such as *"FRECKLES"* may be entered as a subfield consisting of two information items. The first is *"PHYSICAL"* followed by the *"US"* separator, followed by the characteristic as listed in Part 4 Section 13 of the Eighth (or current) Edition of the NCIC Code Manual, July 14, 1999. The *"OTHER"* category is used to enter unlisted or miscellaneous attributes of the facial image. This information shall be entered as a two-information item subfield. The first is *"OTHER"* followed by the *"US"* separator, followed by the unformatted text used to describe the attribute. Multiple attributes and subfields may be listed but must be separated by the *"RS"* character.

17.1.17 Field 10.023-039: Reserved for future definition (RSV)

These fields are reserved for inclusion in future revisions of this standard. None of these fields are to be used at this revision level. If any of these fields are present, they are to be ignored.

17.1.18 Field 10.040: NCIC designation code (SMT)

This field is mandatory for a Type-10 record containing SMT image data. It is used to identify a general location of the captured scar, mark, or tattoo image. The contents of this field will be an entry chosen from Part 4 Section 13 of the Eighth (or current) Edition of the NCIC Code Manual, July 14, 1999. The captured image can encompass an area larger than that specified by a single NCIC body part code for the particular image type. This situation can be accommodated by listing multiple NCIC codes separated by the *"RS"* separator character. In this case the primary code is listed first.

For the *"marks"* category, the NCIC manual lists the common locations for needle track marks. For other body part locations not listed under the *"marks"* category, use the body location codes listed for scars

17.1.19 Field 10.041: SMT size (SMS)

This optional field shall contain the dimensions of the scar, mark or tattoo. It shall consist of two information items. The height shall be the first information item followed by the *"US"* separator character followed by the width. Each dimension shall be entered to the nearest centimeter.

17.1.20 Field 10.042: SMT descriptors (SMD)

This optional field is used to describe the content of the SMT image. It shall consist of one or more subfields. Each subfield shall contain three or four information items that provide progressively detailed information describing the total image or a portion of the image.

The first information item of each subfield shall identify the source of the image as being a scar, a mark, or a tattoo. It shall contain *"SCAR"* to indicate healed scar tissue that was the result an accident or medical procedure. An entry of *"MARK"* shall be used for the pattern resulting from needle or "Track" marks. For either case the second and third information items shall contain "OTHER" and "MISC" and the fourth information item shall contain a textual description or other information concerning the scar or mark pattern.

For deliberately applied or drawn images, the first information item will contain *"TATTOO"* to indicate a common tattoo or indelible image resulting from the pricking of the skin with a coloring matter; *"CHEMICAL"* if the image was created by the use of chemicals to burn the image into the skin; *"BRANDED"* if the image was burned into the skin using a branding iron or other form of heat; or *"CUT"* if the image was caused by incision of the skin.

The second information item shall be the general class code of tattoo chosen from Table 12. For each general class of tattoo, there are several defined subclasses. The third information item of the subfield shall be the appropriate subclass code selected from Tables 13a - 13h which lists the various subclasses of tattoos for each of the general classes.

The final and optional information item in this subfield shall be an ASCII text string that provides additional qualifiers to describe the image or portion of the image. For example, to fully describe a tattoo, there may be a class description of *"ANIMAL"*, with a subclass description of *"DOG"*, and qualified by *"golden retriever with an overbite"*. The *"US"* separator character will be used between information items.

An SMT image consisting of several parts or sub-images shall use multiple subfields, separated by the *"RS"* separator, to fully describe the various parts or features found in the total image. The first subfield shall describe the most predominant feature or sub-image contained in the SMT image. Subsequent subfields shall describe additional portions of the image that are not part of the main or central focal point of the image. For example, a tattoo consisting of a man with a snake on the arm being followed by a dog may contain three subfields - one describing the man, a second describing the snake, and a third describing the dog.

17.1.21 Field 10.043: Color (COL)

This optional field shall contain one subfield corresponding to each subfield contained in Field 10.042. Each subfield shall contain one or more information items that list the color(s) of the tattoo or part of the tattoo. For each subfield, the first information item in the subfield shall be the predominant color chosen from Table 14. Additional colors for the sub-field shall be entered as information items in the subfield separated by the *US* separator character.

17.1.22 Field 10.044-199: Reserved for future definition (RSV)

These fields are reserved for inclusion in future revisions of this standard. None of these fields are to be used at this revision level. If any of these fields are present, they are to be ignored.

17.1.23 Field 10.200-998: User-defined fields (UDF)

These fields are user-definable fields. Their size and content shall be defined by the user and be in accordance with the receiving agency. If present they shall contain ASCII textual information.

17.1.24 Field 10.999: Image data (DAT)

This field shall contain all of the grayscale or color data from a face, scar, mark, tattoo, or other image. It shall always be assigned field number 999 and must be the last physical field in the record. For example, "10.999:" is followed by image data in a binary representation.

Table 12 – Tattoo classes

Class description	Class code
Human Forms and Features	HUMAN
Animals and Animal Features	ANIMAL
Plants	PLANT
Flags	FLAG
Objects	OBJECT
Abstractions	ABSTRACT
Insignias & Symbols	SYMBOL
Other Images	OTHER

Table 13a – Human tattoo subclasses

Subclass	Subclass code
Male Face	MFACE
Female Face	FFACE
Abstract Face	ABFACE
Male Body	MBODY
Female Body	FBODY
Abstract Body	ABBODY
Roles (Knight, Witch, man, etc.)	ROLES
Sports Figures (Football Player, Skier, etc.)	SPORT
Male Body Parts	MBPART
Female Body Parts	FBPART
Abstract Body Parts	ABBPART
Skulls	SKULL
Miscellaneous Human Forms	MHUMAN

Table 13b – Animal tattoo subclasses

Subclass	Subclass code
Cats & Cat Heads	CAT
Dogs & Dog Heads	DOG
Other Domestic Animals	DOMESTIC
Vicious Animals (Lions, Tigers, etc.)	VICIOUS
Horses (Donkeys, Mules, etc.)	HORSE
Other Wild Animals	WILD
Snakes	SNAKE
Dragons	DRAGON
Birds (Cardinal, Hawk, etc.)	BIRD
Spiders, Bugs, and Insects	INSECT
Abstract Animals	ABSTRACT
Animal Parts	PARTS
Miscellaneous Animal Forms	MANIMAL

Table 13c – Plant tattoo subclasses

Subclass	Subclass code
Narcotics	NARCOTICS
Red Flowers	REDFL
Blue Flowers	BLUEFL
Yellow Flowers	YELFL
Drawings of Flowers	DRAW
Rose	ROSE
Tulip	TULIP
Lily	LILY
Miscellaneous Plants, Flowers, Vegetables	MPLANT

Table 13d – Flags tattoo subclasses

Subclass	Subclass code
American Flag	USA
State Flag	STATE
Nazi Flag	NAZI
Confederate Flag	CONFED
British Flag	BRIT
Miscellaneous Flags	MFLAG

Table 13e – Objects tattoo subclasses

Subclass	Subclass code
Fire	FIRE
Weapons (Guns, Arrows, etc.)	WEAP
Airplanes	PLANE
Boats, Ships, and Other Vessels	VESSEL
Trains	TRAIN
Cars, Trucks, and Vehicles	VEHICLE
Mythical (Unicorns, etc.)	MYTH
Sporting Objects (Football, Ski, Hurdles, etc.)	SPORT
Water & Nature Scenes (Rivers, Sky, Trees, etc.)	NATURE
Miscellaneous Objects	MOBJECTS

Table 13f – Abstract tattoo subclasses

Subclass	Subclass code
Figure(s)	FIGURE
Sleeve	SLEEVE
Bracelet	BRACE
Anklet	ANKLET
Necklace	NECKLC
Shirt	SHIRT
Body Band	BODBND
Head Band	HEDBND
Miscellaneous Abstract	MABSTRACT

Table 13g – Symbols tattoo subclasses

Subclass	Subclass code
National Symbols	NATION
Political Symbols	POLITIC
Military Symbols	MILITARY
Fraternal Symbols	FRATERNAL
Professional Symbols	PROFESS
Gang Symbols	GANG
Miscellaneous Symbols	MSYMBOLS

Table 13h – Other tattoo subclasses

Subclass	Subclass code
Wording (Mom, Dad, Mary, etc.)	WORDING
Freeform Drawings	FREEFRM
Miscellaneous Images	MISC

Table 14 – Color codes

Color description	Color code
Black	BLACK
Brown	BROWN
Gray	GRAY
Blue	BLUE
Green	GREEN
Orange	ORANGE
Purple	PURPLE
Red	RED
Yellow	YELLOW
White	WHITE
Multi-colored	MULTI
Outlined	OUTLINE

Each pixel of uncompressed grayscale data shall be quantized to eight bits (256 gray levels) and shall occupy a single byte. Uncompressed color image data shall be expressed as 24 bit RGB pixels. The first byte shall contain the eight bits for the red component of the pixel, the second byte shall contain the eight bits for the green component of the pixel, and the third byte shall contain the last eight bits for the blue component of the pixel. If compression is used, the pixel data shall be compressed in accordance with the compression technique specified in the GCA field. If the JPEG algorithm is to be used to compress the data, this field shall be encoded using the JFIF format specification.

17.2 End of Type-10 logical record

For the sake of consistency, immediately following the last byte of data from field 10.999 an *"FS"* separator shall be used to separate it from the next logical record. This separator must be included in the length field of the Type-10 record.

17.3 Additional facial & SMT image records

Additional Type-10 records may be included in the file. For each additional facial or SMT image, a complete Type-10 logical record together with the *"FS"* separator is required.

18 Type-11 record reserved for future use

19 Type-12 record reserved for future use

20 Type-13 variable-resolution latent image record

The Type-13 tagged-field logical record shall contain image data acquired from latent images. These images are intended to be transmitted to agencies that will automatically extract or provide human intervention and processing to extract the desired feature information from the images.

Information regarding the scanning resolution used, the image size, and other parameters required to process the image, are recorded as tagged-fields within the record.

Table 15 – Type-13 variable-resolution latent record layout

Ident	Cond code	Field number	Field name	Char type	Field size per occurrence		Occur count		Max byte count
					Min	max	min	Max	
LEN	M	13.001	LOGICAL RECORD LENGTH	N	4	8	1	1	15
IDC	M	13.002	IMAGE DESIGNATION CHARACTER	N	2	5	1	1	12
IMP	M	13.003	IMPRESSION TYPE	A	2	2	1	1	9
SRC	M	13.004	SOURCE AGENCY / ORI	AN	10	21	1	1	28
LCD	M	13.005	LATENT CAPTURE DATE	N	9	9	1	1	16
HLL	M	13.006	HORIZONTAL LINE LENGTH	N	4	5	1	1	12
VLL	M	13.007	VERTICAL LINE LENGTH	N	4	5	1	1	12
SLC	M	13.008	SCALE UNITS	N	2	2	1	1	9
HPS	M	13.009	HORIZONTAL PIXEL SCALE	N	2	5	1	1	12
VPS	M	13.010	VERTICAL PIXEL SCALE	N	2	5	1	1	12
CGA	M	13.011	COMPRESSION ALGORITHM	A	5	7	1	1	14
BPX	M	13.012	BITS PER PIXEL	N	2	3	1	1	10
FGP	M	13.013	FINGER POSITION	N	2	3	1	6	25
RSV		13.014 13.019	RESERVED FOR FUTURE DEFINITION	--	--	--	--	--	--
COM	O	13.020	COMMENT	A	2	128	0	1	128
RSV		13.021 13.199	RESERVED FOR FUTURE DEFINITION	--	--	--	--	--	--
UDF	O	13.200 13.998	USER-DEFINED FIELDS	--	--	--	--	--	--
DAT	M	13.999	IMAGE DATA	B	2	--	1	1	--

Key for character type: N = Numeric; A = Alphabetic; AN = Alphanumeric; B = Binary

20.1 Fields for the Type-13 logical record

The following paragraphs describe the data contained in each of the fields for the Type-13 logical record.

Within a Type-13 logical record, entries shall be provided in numbered fields. It is required that the first two fields of the record are ordered, and the field containing the image data shall be the last physical field in the record. For each field of the Type-13 record, Table 15 lists the "condition code" as being mandatory "M" or optional "O", the field number, the field name, character type, field size, and occurrence limits. Based on a three digit field number, the maximum byte count size for the field is given in the last column. As more digits are used for the field number, the maximum byte count will also increase. The two entries in the "field size per occurrence" include all character separators used in the field. The "maximum byte count" includes the field number, the information, and all the character separators including the "GS" character.

20.1.1 Field 13.001: Logical record length (LEN)

This mandatory ASCII field shall contain the total count of the number of bytes in the Type-13 logical record. Field 13.001 shall specify the length of the record including every character of every field contained in the record and the information separators.

20.1.2 Field 13.002: Image designation character (IDC)

This mandatory ASCII field shall be used to identify the latent image data contained in the record. This IDC shall match the IDC found in the file content (CNT) field of the Type-1 record.

20.1.3 Field 13.003: Impression type (IMP)

This mandatory one- or two-byte ASCII field shall indicate the manner by which the latent image information was obtained. The appropriate latent code choice selected from Table 5 (finger) or Table 18 (palm) shall be entered in this field.

20.1.4 Field 13.004: Source agency / ORI (SRC)

This mandatory ASCII field shall contain the identification of the administration or organization that originally captured the latent image contained in the record. Normally, the ORI of the agency that captured the image will be contained in this field. The SRC may contain up to 20 identifying characters and the data content of this field shall be defined by the user and be in accordance with the receiving agency.

20.1.5 Field 13.005: Latent capture date (LCD)

This mandatory ASCII field shall contain the date that the latent image contained in the record was captured. The date shall appear as eight digits in the format *CCYYMMDD*. The *CCYY* characters shall represent the year the image was captured; the *MM* characters shall be the tens and units values of the month; and the *DD* characters shall be the tens and units values of the day in the month. For example, 20000229 represents February 29, 2000. The complete date must be a legitimate date.

20.1.6 Field 13.006: Horizontal line length (HLL)

This mandatory ASCII field shall contain the number of pixels contained on a single horizontal line of the transmitted image.

20.1.7 Field 13.007: Vertical line length (VLL)

This mandatory ASCII field shall contain the number of horizontal lines contained in the transmitted image.

20.1.8 Field 13.008: Scale units (SLC)

This mandatory ASCII field shall specify the units used to describe the image sampling frequency (pixel density). A "1" in this field indicates pixels per inch, or a "2" indicates pixels per centimeter. A "0" in this field indicates no scale is given. For this case, the quotient of HPS/VPS gives the pixel aspect ratio.

20.1.9 Field 13.009: Horizontal pixel scale (HPS)

This mandatory ASCII field shall specify the integer pixel density used in the horizontal direction providing the SLC contains a "1" or a "2". Otherwise, it indicates the horizontal component of the pixel aspect ratio.

20.1.10 Field 13.010: Vertical pixel scale (VPS)

This mandatory ASCII field shall specify the integer pixel density used in the vertical direction providing the SLC contains a "1" or a "2". Otherwise, it indicates the vertical component of the pixel aspect ratio.

20.1.11 Field 13.011: Compression algorithm (CGA)

This mandatory ASCII field shall specify the algorithm used to compress grayscale images. An entry of *"NONE"* in this field indicates that the data contained in this record is uncompressed. For those images that are to be losslessly compressed, this field shall contain the preferred method for the compression of latent fingerprint images. For grayscale images, the domain registrar shall maintain a registry of compression techniques and corresponding codes that may be used as they become available.

20.1.12 Field 13.012: Bits per pixel (BPX)

This mandatory ASCII field shall contain the number of bits used to represent a pixel. This field shall contain an entry of "8" for normal grayscale values of "0" to "255". Any entry in this field greater than "8" shall represent a grayscale pixel with increased precision.

20.1.13 Field 13.013: Finger / palm position (FGP)

This mandatory tagged-field shall contain one or more the possible finger or palm positions that may match the latent image. The decimal code number corresponding to the known or most probable finger position shall be taken from Table 6 or the most probable palm position from Table 19 and entered as a one- or two-character ASCII subfield. Additional finger and/or palm positions may be referenced by entering the alternate position codes as subfields separated by the *"RS"* separator character. The code "0", for "Unknown Finger", shall be used to reference every finger position from one through ten. The code "20", for

"Unknown Palm", shall be used to reference every listed palmprint position.

20.1.14 Field 13.014-019: Reserved for future definition (RSV)

These fields are reserved for inclusion in future revisions of this standard. None of these fields are to be used at this revision level. If any of these fields are present, they are to be ignored.

20.1.15 Field 13.020: Comment (COM)

This optional field may be used to insert comments or other ASCII text information with the latent image data.

20.1.16 Field 13.021-199: Reserved for future definition (RSV)

These fields are reserved for inclusion in future revisions of this standard. None of these fields are to be used at this revision level. If any of these fields are present, they are to be ignored.

20.1.17 Fields 13.200-998: User-defined fields (UDF)

These fields are user-definable fields. Their size and content shall be defined by the user and be in accordance with the receiving agency. If present they shall contain ASCII textual information.

20.1.18 Field 13.999: Image data (DAT)

This field shall contain all of data from a captured latent image. It shall always be assigned field number 999 and must be the last physical field in the record. For example, "13.999:" is followed by image data in a binary representation.

Each pixel of uncompressed grayscale data shall normally be quantized to eight bits (256 gray levels) contained in a single byte. If the entry in BPX Field 13.012 is greater or less than "8", the number of bytes required to contain a pixel will be different. If compression is used, the pixel data shall be compressed in accordance with the compression technique specified in the GCA field.

20.2 End of Type-13 variable-resolution latent image record

For the sake of consistency, immediately following the last byte of data from field 13.999 an *"FS"* separator shall be used to separate it from the next logical record. This separator must be included in the length field of the Type-13 record.

20.3 Additional variable-resolution latent image records

Additional Type-13 records may be included in the file. For each additional latent image, a complete Type-13 logical record together with the *"FS"* separator is required.

21 Type-14 variable-resolution tenprint image record

The Type-14 tagged-field logical record shall contain and be used to exchange tenprint fingerprint image data. Rolled and plain fingerprint impressions shall be acquired from a tenprint card or from a live-scan device. Captured images are intended to be transmitted to agencies that will automatically extract the desired feature information from the images for matching purposes.

Textual information regarding the scanning resolution used, the image size and other parameters or comments required to process the image are recorded as tagged-fields within the record.

21.1 Fields for the Type-14 logical record

The following paragraphs describe the data contained in each of the fields for the Type-14 logical record.

Within a Type-14 logical record, entries shall be provided in numbered fields. It is required that the first two fields of the record are ordered, and the field containing the image data shall be the last physical field in the record. For each field of the Type-14 record, Table 16 lists the "condition code" as being mandatory "M" or optional "O", the field number, the field name, character type, field size, and occurrence limits. Based on a three digit field number, the maximum byte count size for the field is given in the last column. As more digits are used for the field number, the maximum

Table 16 – Type-14 variable-resolution tenprint record layout

Ident	Cond code	Field number	Field name	Char type	Field size per occurrence		Occur count		Max byte count
					min	max	min	Max	
LEN	M	14.001	LOGICAL RECORD LENGTH	N	4	8	1	1	15
IDC	M	14.002	IMAGE DESIGNATION CHARACTER	N	2	5	1	1	12
IMP	M	14.003	IMPRESSION TYPE	A	2	2	1	1	9
SRC	M	14.004	SOURCE AGENCY / ORI	AN	10	21	1	1	28
TCD	M	14.005	TENPRINT CAPTURE DATE	N	9	9	1	1	16
HLL	M	14.006	HORIZONTAL LINE LENGTH	N	4	5	1	1	12
VLL	M	14.007	VERTICAL LINE LENGTH	N	4	5	1	1	12
SLC	M	14.008	SCALE UNITS	N	2	2	1	1	9
HPS	M	14.009	HORIZONTAL PIXEL SCALE	N	2	5	1	1	12
VPS	M	14.010	VERTICAL PIXEL SCALE	N	2	5	1	1	12
CGA	M	14.011	COMPRESSION ALGORITHM	A	5	7	1	1	14
BPX	M	14.012	BITS PER PIXEL	N	2	3	1	1	10
FGP	M	14.013	FINGER POSITION	N	2	3	1	6	25
RSV		14.014 14.019	RESERVED FOR FUTURE DEFINITION	--	--	--	--	--	--
COM	O	14.020	COMMENT	A	2	128	0	1	128
RSV		14.021 14.199	RESERVED FOR FUTURE DEFINITION	--	--	--	--	--	--
UDF	O	14.200 14.998	USER-DEFINED FIELDS	--	--	--	--	--	--
DAT	M	14.999	IMAGE DATA	B	2	--	1	1	--

Key for character type: N = Numeric; A = Alphabetic; AN = Alphanumeric; B = Binary

byte count will also increase. The two entries in the "field size per occurrence" include all character separators used in the field. The "maximum byte count" includes the field number, the information, and all the character separators including the *"GS"* character.

21.1.1 Field 14.001: Logical record length (LEN)

This mandatory ASCII field shall contain the total count of the number of bytes in the Type-14 logical record. Field 14.001 shall specify the length of the record including every character of every field contained in the record and the information separators.

21.1.2 Field 14.002: Image designation character (IDC)

This mandatory ASCII field shall be used to identify the tenprint fingerprint image contained in the record. This IDC shall match the IDC found in the file content (CNT) field of the Type-1 record.

21.1.3 Field 14.003: Impression type (IMP)

This mandatory one-byte ASCII field shall indicate the manner by which the tenprint image information was obtained. The appropriate code selected from Table 5 shall be entered in this field.

21.1.4 Field 14.004: Source agency / ORI (SRC)

This mandatory ASCII field shall contain the identification of the administration or organization that originally captured the tenprint image contained in the record. Normally, the ORI of the agency that captured the image will be contained in this field. The SRC may contain up to 20 identifying characters and the data content of this field shall be defined by the user and be in accordance with the receiving agency.

21.1.5 Field 14.005: Tenprint capture date (TCD)

This mandatory ASCII field shall contain the date that the tenprint image was captured. The date shall appear as eight digits in the format CCYYMMDD. The CCYY characters shall represent the year the image was captured; the MM characters shall be the tens and units values of the month; and the DD characters shall be the tens and units values of the day in the month. For example, the entry 20000229 represents February 29, 2000. The complete date must be a legitimate date.

21.1.6 Field 14.006: Horizontal line length (HLL)

This mandatory ASCII field shall contain the number of pixels contained on a single horizontal line of the transmitted image.

21.1.7 Field 14.007: Vertical line length (VLL)

This mandatory ASCII field shall contain the number of horizontal lines contained in the transmitted image.

21.1.8 Field 14.008: Scale units (SLC)

This mandatory ASCII field shall specify the units used to describe the image sampling frequency (pixel density). A "1" in this field indicates pixels per inch, or a "2" indicates pixels per centimeter. A "0" in this field indicates no scale is given. For this case, the quotient of HPS/VPS gives the pixel aspect ratio.

21.1.9 Field 14.009: Horizontal pixel scale (HPS)

This mandatory ASCII field shall specify the integer pixel density used in the horizontal direction providing the SLC contains a "1" or a "2". Otherwise, it indicates the horizontal component of the pixel aspect ratio.

21.1.10 Field 14.010: Vertical pixel scale (VPS)

This mandatory ASCII field shall specify the integer pixel density used in the vertical direction providing the SLC contains a "1" or a "2". Otherwise, it indicates the vertical component of the pixel aspect ratio.

21.1.11 Field 14.011: Compression algorithm (CGA)

This mandatory ASCII field shall specify the algorithm used to compress grayscale images. An entry of "NONE" in this field indicates that the data contained in this record is uncompressed. For those images that are to be compressed, this field shall contain the preferred method for the compression of tenprint fingerprint images. For grayscale images, the domain registrar maintains a registry of acceptable compression techniques and corresponding codes that may be used as they become available.

21.1.12 Field 14.012: Bits per pixel (BPX)

This mandatory ASCII field shall contain the number of bits used to represent a pixel. This field shall contain an entry of "8" for normal grayscale values of "0" to "255". Any entry in this field greater than or less than "8" shall represent a grayscale pixel with increased or decreased precision respectively.

21.1.13 Field 14.013: Finger position (FGP)

This mandatory tagged-field shall contain finger position that matches the tenprint image. The decimal code number corresponding to the known or most probable finger position shall be taken from Table 6 and entered as a one- or two-character ASCII subfield. Table 6 also lists the maximum image area that can be transmitted for each of the fourteen possible finger positions. Additional finger positions may be referenced in the transaction by entering the alternate finger positions as subfields separated by the *"RS"* separator character. The code "0", for "Unknown Finger", shall be used to reference every finger position from one through ten.

21.1.14 Field 14.014-019: Reserved for future definition (RSV)

These fields are reserved for inclusion in future revisions of this standard. None of these fields are to be used at this revision level. If any of these fields are present, they are to be ignored.

21.1.15 Field 14.020: Comment (COM)

This optional field may be used to insert comments or other ASCII text information with the tenprint image data.

21.1.16 Field 14.021-199: Reserved for future definition (RSV)

These fields are reserved for inclusion in future revisions of this standard. None of these fields are to be used at this revision level. If any of these fields are present, they are to be ignored.

21.1.17 Fields 14.200-998: User-defined fields (UDF)

These fields are user-definable fields. Their size and content shall be defined by the user and be in accordance with the receiving agency. If present they shall contain ASCII textual information.

21.1.18 Field 14.999: Image data (DAT)

This field shall contain all of the data from a captured tenprint image. It shall always be assigned field number 999 and must be the last physical field in the record. For example, "14.999:" is followed by image data in a binary representation.

Each pixel of uncompressed grayscale data shall normally be quantized to eight bits (256 gray levels) contained in a single byte. If the entry in BPX Field 14.012 is greater or less than "8", the number of bytes required to contain a pixel will be different. If compression is used, the pixel data shall be compressed in accordance with the compression technique specified in the CGA field.

21.2 End of Type-14 variable-resolution tenprint image record

For the sake of consistency, immediately following the last byte of data from field 14.999 an *"FS"* separator shall be used to separate it from the next logical record. This separator must be included in the length field of the Type-14 record.

21.3 Additional variable-resolution tenprint image records

Additional Type-14 records may be included in the file. For each additional tenprint image, a complete Type-14 logical record together with the *"FS"* separator is required.

22 Type-15 variable-resolution palmprint image record

The Type-15 tagged-field logical record shall contain and be used to exchange palmprint image data together with fixed and user-defined textual information fields pertinent to the digitized image. Information regarding the scanning resolution used, the image size and other parameters or comments required to process the image are recorded as tagged-fields within the record. Palmprint images transmitted to other agencies will be processed by the recipient agencies to extract the desired feature information required for matching purposes.

The image data shall be acquired directly from a subject using a live-scan device, or from a palmprint card or other media that contains the subject's palmprints.

Any method used to acquire the palmprint images shall be capable of capturing a set of images for each hand. This set shall include the writer's palm as a single scanned image, and the entire area of the full palm extending from the wrist bracelet to the tips of the fingers as one or two scanned images. If two images are used to represent the full palm, the lower image shall extend from the wrist bracelet to the top of the interdigital area (third finger joint) and shall include the thenar, and hypothenar areas of the palm. The upper image shall extend from the bottom of the interdigital area to the upper tips of the fingers. This provides an adequate amount of overlap between the two images that are both located over the interdigital area of the palm. By matching the ridge structure and details contained in this common area, an examiner can confidently state that both images came from the same palm.

As a palmprint transaction may be used for different purposes, it may contain one or more unique image areas recorded from the palm or hand. A complete palmprint record set for one individual will normally include the writer's palm and the full palm image(s) from each hand. Since a tagged-field logical image record may contain only one binary field, a single Type-15 record will be required for each writer's palm and one or two Type-15 records for each full palm. Therefore, four to six Type-15 records will be required to represent the subject's palmprints in a normal palmprint transaction.

22.1 Fields for the Type-15 logical record

The following paragraphs describe the data contained in each of the fields for the Type-15 logical record.

Within a Type-15 logical record, entries shall be provided in numbered fields. It is required that the first two fields of the record are ordered, and the field containing the image data shall be the last physical field in the record. For each field of the Type-15 record, Table 17 lists the "condition code" as being mandatory "M" or optional "O", the field number, the field name, character type, field size, and occurrence limits. Based on a three digit field number, the maximum byte count size for the field is given in the last column. As more digits are used for the field number, the maximum byte count will also increase. The two entries in the "field size per occurrence" include all character separators used in the field. The "maximum byte count" includes the field number, the information, and all the character separators including the *"GS"* character.

22.1.1 Field 15.001: Logical record length (LEN)

This mandatory ASCII field shall contain the total count of the number of bytes in the Type-15 logical record. Field 15.001 shall specify the length of the record including every character of every field contained in the record and the information separators.

22.1.2 Field 15.002: Image designation character (IDC)

This mandatory ASCII field shall be used to identify the palmprint image contained in the record. This IDC shall match the IDC found in the file content (CNT) field of the Type-1 record.

22.1.3 Field 15.003: Impression type (IMP)

This mandatory one-byte ASCII field shall indicate the manner by which the palmprint image information was obtained. The appropriate code selected from Table 18 shall be entered in this field.

22.1.4 Field 15.004: Source agency/ORI (SRC)

This mandatory ASCII field shall contain the identification of the administration or organization that originally captured the palmprint image contained in the record. Normally, the ORI of the agency that captured the image will be contained in this field. The SRC may contain up to 20 identifying characters and the data content of this field shall be defined by the user and be in accordance with the receiving agency.

22.1.5 Field 15.005: Palmprint capture date (PCD)

This mandatory ASCII field shall contain the date that the palmprint image was captured. The date shall appear as eight digits in the format CCYYMMDD. The CCYY characters shall represent the year the image was captured; the MM characters shall be the tens and units values of the month; and the DD characters shall be the tens and units values of the day in the month. For example, the entry 20000229 represents February 29, 2000. The complete date must be a legitimate date.

22.1.6 Field 15.006: Horizontal line length (HLL)

This mandatory ASCII field shall contain the number of pixels contained on a single horizontal line of the transmitted image.

22.1.7 Field 15.007: Vertical line length (VLL)

This mandatory ASCII field shall contain the number of horizontal lines contained in the transmitted image.

22.1.8 Field 15.008: Scale units (SLC)

This mandatory ASCII field shall specify the units used to describe the image sampling frequency (pixel density). A "1" in this field indicates pixels per inch, or a "2" indicates pixels per centimeter. A "0" in this field indicates no scale is given. For this case, the quotient of HPS/VPS gives the pixel aspect ratio.

22.1.9 Field 15.009: Horizontal pixel scale (HPS)

This mandatory ASCII field shall specify the integer pixel density used in the horizontal direction providing the SLC contains a "1" or a "2". Otherwise, it indicates the horizontal component of the pixel aspect ratio.

22.1.10 Field 15.010: Vertical pixel scale (VPS)

This mandatory ASCII field shall specify the integer pixel density used in the vertical direction providing the SLC contains a "1" or a "2". Otherwise,

ANSI/NIST-ITL 1-2000

Table 17 – Type-15 variable-resolution palmprint record layout

Ident	Cond code	Field number	Field name	Char type	Field size per occurrence		Occur count		Max byte count
					Min	max	min	max	
LEN	M	15.001	LOGICAL RECORD LENGTH	N	4	8	1	1	15
IDC	M	15.002	IMAGE DESIGNATION CHARACTER	N	2	5	1	1	12
IMP	M	15.003	IMPRESSION TYPE	N	2	2	1	1	9
SRC	M	15.004	SOURCE AGENCY / ORI	AN	10	21	1	1	28
PCD	M	15.005	PALMPRINT CAPTURE DATE	N	9	9	1	1	16
HLL	M	15.006	HORIZONTAL LINE LENGTH	N	4	5	1	1	12
VLL	M	15.007	VERTICAL LINE LENGTH	N	4	5	1	1	12
SLC	M	15.008	SCALE UNITS	N	2	2	1	1	9
HPS	M	15.009	HORIZONTAL PIXEL SCALE	N	2	5	1	1	12
VPS	M	15.010	VERTICAL PIXEL SCALE	N	2	5	1	1	12
CGA	M	15.011	COMPRESSION ALGORITHM	AN	5	7	1	1	14
BPX	M	15.012	BITS PER PIXEL	N	2	3	1	1	10
PLP	M	15.013	PALMPRINT POSITION	N	2	3	1	1	10
RSV		15.014 15.019	RESERVED FOR FUTURE INCLUSION	--	--	--	--	--	--
COM	O	15.020	COMMENT	AN	2	128	0	1	128
RSV		15.021 15.199	RESERVED FOR FUTURE INCLUSION	--	--	--	--	--	--
UDF	O	15.200 15.998	USER-DEFINED FIELDS	--	--	--	--	--	--
DAT	M	15.999	IMAGE DATA	B	2	--	1	1	--

Key for character type: N = Numeric; A = Alphabetic; AN = Alphanumeric; B = Binary

Table 18 – Palm impression type

Description	Code
Live-scan palm	10
Nonlive-scan palm	11
Latent palm impression	12
Latent palm tracing	13
Latent palm photo	14
Latent palm lift	15

it indicates the vertical component of the pixel aspect ratio.

22.1.11 Field 15.011: Compression algorithm (CGA)

This mandatory ASCII field shall specify the algorithm used to compress grayscale images. An entry of "NONE" in this field indicates that the data contained in this record is uncompressed. For those images that are to be compressed, this field shall contain the preferred method for the compression of tenprint fingerprint images. For grayscale images, the domain registrar maintains a registry of acceptable compression techniques and corresponding codes that may be used as they become available.

22.1.12 Field 15.012: Bits per pixel (BPX)

This mandatory ASCII field shall contain the number of bits used to represent a pixel. This field shall contain an entry of "8" for normal grayscale values of "0" to "255". Any entry in this field greater than or less than "8" shall represent a grayscale pixel with increased or decreased precision respectively.

Table 19 – Palm codes, areas, & sizes

Palm Position	Palm code	Image area (mm^2)	Width (mm)	(in)	Height (mm)	(in)
Unknown Palm	20	28387	139.7	5.5	203.2	8.0
Right Full Palm	21	28387	139.7	5.5	203.2	8.0
Right Writer's Palm	22	5645	44.5	1.8	127.0	5.0
Left Full Palm	23	28387	139.7	5.5	203.2	8.0
Left Writer's Palm	24	5645	44.5	1.8	127.0	5.0
Right Lower Palm	25	19516	139.7	5.5	139.7	5.5
Right Upper Palm	26	19516	139.7	5.5	139.7	5.5
Left Lower Palm	27	19516	139.7	5.5	139.7	5.5
Left Upper Palm	28	19516	139.7	5.5	139.7	5.5
Right Other	29	28387	139.7	5.5	203.2	8.0
Left Other	30	28387	139.7	5.5	203.2	8.0

22.1.13 Field 15.013: Palmprint position (PLP)

This mandatory tagged-field shall contain the palmprint position that matches the palmprint image. The decimal code number corresponding to the known or most probable palmprint position shall be taken from Table 19 and entered as a two-character ASCII subfield. Table 19 also lists the maximum image areas and dimensions for each of the possible palmprint positions.

22.1.14 Field 15.014-019: Reserved for future definition (RSV)

These fields are reserved for inclusion in future revisions of this standard. None of these fields are to be used at this revision level. If any of these fields are present, they are to be ignored.

22.1.15 Field 15.020: Comment (COM)

This optional field may be used to insert comments or other ASCII text information with the palmprint image data.

22.1.16 Field 15.021-199: Reserved for future definition (RSV)

These fields are reserved for inclusion in future revisions of this standard. None of these fields are to be used at this revision level. If any of these fields are present, they are to be ignored.

22.1.17 Fields 15.200-998: User-defined fields (UDF)

These fields are user-definable fields. Their size and content shall be defined by the user and be in accordance with the receiving agency. If present they shall contain ASCII textual information.

22.1.18 Field 15.999: Image data (DAT)

This field shall contain all of the data from a captured palmprint image. It shall always be assigned field number 999 and must be the last physical field in the record. For example, "15.999:" is followed by image data in a binary representation. Each pixel of uncompressed grayscale data shall normally be quantized to eight bits (256 gray levels) contained in a single byte. If the entry in BPX Field 15.012 is greater or less than 8, the number of bytes required to contain a pixel will be different. If compression is used, the pixel data shall be compressed in accordance with the compression technique specified in the CGA field.

22.2 End of Type-15 variable-resolution palmprint image record

For the sake of consistency, immediately following the last byte of data from field 15.999 an *"FS"* separator shall be used to separate it from the next logical record. This separator must be included in the length field of the Type-15 record.

Table 20 – Type-16 User-defined testing record layout

Ident	Cond code	Field number	Field name	Char type	Field size per occurrence		Occur count		Max byte count
					Min	max	min	Max	
LEN	M	16.001	LOGICAL RECORD LENGTH	N	4	8	1	1	15
IDC	M	16.002	IMAGE DESIGNATION CHARACTER	N	2	5	1	1	12
UDF	O	16.003 16.005	USER-DEFINED FIELDS	--	--	--	--	--	--
HLL	M	16.006	HORIZONTAL LINE LENGTH	N	4	5	1	1	12
VLL	M	16.007	VERTICAL LINE LENGTH	N	4	5	1	1	12
SLC	M	16.008	SCALE UNITS	N	2	2	1	1	9
HPS	M	16.009	HORIZONTAL PIXEL SCALE	N	2	5	1	1	12
VPS	M	16.010	VERTICAL PIXEL SCALE	N	2	5	1	1	12
CGA	M	16.011	COMPRESSION ALGORITHM	AN	5	7	1	1	14
BPX	M	16.012	BITS PER PIXEL	N	2	3	1	1	10
UDF	O	16.013 16.998	USER-DEFINED FIELDS	--	--	--	--	--	--
DAT	M	16.999	IMAGE DATA	B	2	--	1	1	--

Key for character type: N = Numeric; A = Alphabetic; AN = Alphanumeric; B = Binary

22.3 Additional Type-15 variable-resolution palmprint image records

Additional Type-15 records may be included in the file. For each additional palmprint image, a complete Type-15 logical record together with the *"FS"* separator is required.

23 Type-16 user-defined testing image record

The Type-16 tagged-field logical record shall contain and be used to exchange image data together with textual information fields pertinent to the digitized image. This logical record type allows the standard to provide the ability to exchange images not addressed by other record types in the standard. It is intended as the tagged-field user-defined logical record to be used for developmental purposes.

The image data contained in the Type-16 logical record may be in a compressed form. With the exception of the tagged-fields described below, the format, parameters, and types of images to be exchanged are undefined by this Standard and shall be agreed upon between the sender and recipient.

23.1 Fields for the Type-16 logical record

The following paragraphs describe the data contained in each of the fields for the Type-16 logical record.

Within a Type-16 logical record, entries shall be provided in tagged numbered fields as described below. The logical record length, and the IDC must be provided as the first two ordered tagged-fields with the image data contained in the last physical field of the record. Fields describing the physical parameters of the image size and resolution are mandatory and must also be provided. These fields and the remaining user-defined fields may be unordered. For each required field of the Type-16 record, Table 20 lists the "condition code" as being mandatory "M" or optional "O", the field number, the field name, character type, field size, and occurrence limits. Based on a three digit field number, the maximum byte count size for the field is given in the last column. As more digits are used for the field number, the maximum byte count will also increase. The two entries in the "field size per occurrence" include all character separators used in the field. The "maximum byte count" includes the field number, the information, and all the character separators including the *"GS"* character.

23.1.1 Field 16.001: Logical record length (LEN)

This mandatory ASCII field shall contain the total count of the number of bytes in the Type-16 logical record. Field 16.001 shall specify the length of the record including every character of every field contained in the record and the information separators.

23.1.2 Field 16.002: Image designation character (IDC)

This mandatory ASCII field shall be used to identify the image contained in the record. This IDC shall match the IDC found in the file content (CNT) field of the Type-1 record.

23.1.3 Fields 16.003-005: User-defined fields (UDF)

These fields are user-definable fields. Their size and content shall be defined by the user and be in accordance with the receiving agency. If present they shall contain ASCII textual information.

23.1.4 Field 16.006: Horizontal line length (HLL)

This mandatory ASCII field shall contain the number of pixels present on a single horizontal line of the transmitted image.

23.1.5 Field 16.007: Vertical line length (VLL)

This mandatory ASCII field shall contain the number of horizontal lines present in the transmitted image.

23.1.6 Field 16.008: Scale units (SLC)

This mandatory ASCII field shall specify the units used to describe the image sampling frequency (pixel density). A "1" in this field indicates pixels per inch, or a "2" indicates pixels per centimeter. A "0" in this field indicates no scale is given. For this case, the quotient of HPS/VPS gives the pixel aspect ratio.

23.1.7 Field 16.009: Horizontal pixel scale (HPS)

This mandatory ASCII field shall specify the integer pixel density used in the horizontal direction providing the SLC contains a "1" or a "2". Otherwise, it indicates the horizontal component of the pixel aspect ratio.

23.1.8 Field 16.010: Vertical pixel scale (VPS)

This mandatory ASCII field shall specify the integer pixel density used in the vertical direction providing the SLC contains a "1" or a "2". Otherwise, it indicates the vertical component of the pixel aspect ratio.

23.1.9 Field 16.011: Compression algorithm (CGA)

This mandatory ASCII field shall specify the algorithm used to compress the image. An entry of "NONE" in this field indicates that the data contained in this record is uncompressed. The domain registrar maintains a registry of acceptable compression techniques and corresponding codes that may be used as they become available.

23.1.10 Field 16.012: Bits per pixel (BPX)

This mandatory ASCII field shall contain the number of bits used to represent a pixel.

23.1.11 Fields 16.013-998: User-defined fields (UDF)

These fields are user-definable fields. Their size and content shall be defined by the user and be in accordance with the receiving agency. If present they shall contain ASCII textual information.

23.1.12 Field 16.999: Image data (DAT)

This field shall contain all of the pixel data from a captured unspecified image. It shall always be assigned field number 999 and must be the last physical field in the record. For example, "16.999:" is followed by image data in a binary representation.

23.2 End of Type-16 user-defined testing image record

For the sake of consistency, immediately following the last byte of data from field 16.999 an *"FS"* separator shall be used to separate it from the next logical record. This separator must be included in the length field of the Type-16 record.

23.3 Additional Type-16 user-defined testing image records

Additional Type-16 records may be included in the file. For each additional image, a complete Type-16 logical record together with the *"FS"* separator is required.

24 Another individual

If fingerprint data for another individual is to be recorded or transmitted, a new file shall be generated for that individual using the same format as described previously.

Annex A - 7-bit ANSI code for information interchange
(normative)

B_7 = MSB →					0	0	0	0	1	1	1	1
b_6 →					0	0	1	1	0	0	1	1
b_5 →					0	1	0	1	0	1	0	1
Bits	b_4 ↓	b_3 ↓	b_2 ↓	b_1 ↓	COLUMN → ROW ↓	0	1	2	3	4	5	6	7
	0	0	0	0	0	NUL	DLE	SP	0	@	P	☐	p
	0	0	0	1	1	SOH	DC1	!	1	A	Q	a	q
	0	0	1	0	2	STX	DC2	"	2	B	R	b	r
	0	0	1	1	3	ETX	DC3	#	3	C	S	c	s
	0	1	0	0	4	EOT	DC4	$	4	D	T	d	t
	0	1	0	1	5	ENQ	NAK	%	5	E	U	e	u
	0	1	1	0	6	ACK	SYN	&	6	F	V	f	v
	0	1	1	1	7	BEL	ETB	☐	7	G	W	g	w
	1	0	0	0	8	BS	CAN	(	8	H	X	h	x
	1	0	0	1	9	HT	EM	)	9	I	Y	i	y
	1	0	1	0	10	LF	SUB	*	:	J	Z	j	z
	1	0	1	1	11	VT	ESC	+	;	K	[	K	{
	1	1	0	0	12	FF	FS	,	<	L	\	☐	|
	1	1	0	1	13	CR	GS	-	=	M	]	m	}
	1	1	1	0	14	SO	RS	.	>	N	^	n	~
	1	1	1	1	15	SI	US	/	?	O	_	o	DEL

Annex B - Use of information separator characters
(informative)

FN is the number of a field (including record type) within a tagged-field record.

IF is the information field associated with an FN.

II is the information item belonging to an IF.

SF is the subfield used for multiple entries of an II or an IF.

$\frac{F}{S}$ File separator character – separates logical records.

$\frac{G}{S}$ Group separator character – separates fields.

$\frac{R}{S}$ Record separator character – separates repeated subfields.

$\frac{U}{S}$ Unit separator character – separates information items.

The $\frac{G}{S}$ is used between fields – the $\frac{F}{S}$ between logical records:

$$FN_j: IF \; \frac{G}{S} \; FN_k : \ldots \frac{F}{S} \; FN_1 : IF \; \frac{G}{S} \ldots \frac{F}{S}$$

For fields with more than one information item, the $\frac{U}{S}$ is used:

$$FN_j : II_a \; \frac{U}{S} \; II_b \; \frac{G}{S} \; FN_k \ldots \frac{F}{S}$$

For fields with multiple subfields, the $\frac{R}{S}$ is used:

$$FN_j : II_a \; \frac{U}{S} \; II_b \; \frac{R}{S} \; II_a \; \frac{U}{S} \; II_b \; \frac{G}{S} \; FN_k \ldots \frac{F}{S}$$

which can be expressed as:

$$FN_j : SF_1 \; \frac{R}{S} \; SF_2 \; \frac{G}{S} \; FN_k \ldots \frac{F}{S}$$

Annex C - Base-64 encoding scheme
(normative)

The base-64 Content-Transfer-Encoding is designed to represent arbitrary sequences of octets in a form that need not be humanly readable. The encoding and decoding algorithms are simple, but the encoded data are consistently only about 33 percent larger than the unencoded data. This encoding is virtually identical to the one used in Privacy Enhanced Mail (PEM) applications, as defined in RFC 1421. The base-64 encoding is adapted from RFC 1421, with one change: base-64 eliminates the "*" mechanism for embedded clear text.

A 65-character subset of US-ASCII is used, enabling 6 bits to be represented per printable character. (The extra 65th character, "=", is used to signify a special processing function.)

NOTE: This subset has the important property that it is represented identically in all versions of ISO 646, including US ASCII and all characters in the subset are also represented identically in all versions of EBCDIC. Other popular encodings, such as the encoding used by the uuencode utility and the base-85 encoding specified as part of Level 2 PostScript, do not share these properties, and thus do not fulfill the portability requirements a binary transport encoding for mail must meet.

The encoding process represents 24-bit groups of input bits as output strings of 4 encoded characters. Proceeding from left to right, concatenating 3 8-bit input groups forms a 24-bit input group. These 24 bits are then treated as 4 concatenated 6-bit groups, each of which is translated into a single digit in the base-64 alphabet. When encoding a bit stream via the base-64 encoding, the bit stream must be presumed to be ordered with the most significant bit first. That is, the first bit in the stream will be the high-order bit in the first byte, and the eighth bit will be the low-order bit in the first byte, and so on.

Each 6-bit group is used as an index into an array of 64 printable characters. The character referenced by the index is placed in the output string. These characters, identified in Table C1, below, are selected so as to be universally representable, and the set excludes characters with particular significance to SMTP (e.g., ".", CR, LF) and to the encapsulation boundaries defined in this document (e.g., "-").

The output stream (encoded bytes) must be represented in lines of no more than 76 characters each. All line breaks or other characters not found in Table C1 must be ignored by decoding software. In base-64 data, characters other than those in Table C1, line breaks, and other white space probably indicate a transmission error, about which a warning message or even a message rejection might be appropriate under some circumstances.

Table C1 – Base-64 alphabet

Value	Encoding	Value	Encoding	Value	Encoding	Value	Encoding
0	A	17	R	34	I	51	z
1	B	18	S	35	j	52	0
2	C	19	T	36	k	53	1
3	D	20	U	37	l	54	2
4	E	21	V	38	m	55	3
5	F	22	W	39	n	56	4
6	G	23	X	40	o	57	5
7	H	24	Y	41	p	58	6
8	I	25	Z	42	q	59	7
9	J	26	a	43	r	60	8
10	K	27	b	44	s	61	9
11	L	28	c	45	t	62	+
12	M	29	d	46	u	63	/
13	N	30	e	47	v		
14	O	31	f	48	w	(pad)	=
15	P	32	g	49	x		
16	Q	33	h	50	y		

Special processing is performed if fewer than 24 bits are available at the end of the data being encoded. A full encoding quantum is always completed at the end of a body. When fewer than 24 input bits are available in an input group, zero bits are added (on the right) to form an integral number of 6-bit groups. Padding at the end of the data is performed using the '=' character. Since all base-64 input is an integral number of octets, only the following cases can arise: (1) the final quantum of encoding input is an integral multiple of 24 bits; here, the final unit of encoded output will be an integral multiple of 4 characters with no "=" padding, (2) the final quantum of encoding input is exactly 8 bits; here, the final unit of encoded output will be two characters followed by two "=" padding characters, or (3) the final quantum of encoding input is exactly 16 bits; here, the final unit of encoded output will be three characters followed by one "=" padding character.

Because it is used only for padding at the end of the data, the occurrence of any '=' characters may be taken as evidence that the end of the data has been reached (without truncation in transit). No such assurance is possible, however, when the number of octets transmitted was a multiple of three.

Any characters outside of the base-64 alphabet are to be ignored in base-64-encoded data. The same applies to any illegal sequence of characters in the base-64 encoding, such as "=====" .

Care must be taken to use the proper octets for line breaks if base-64 encoding is applied directly to text material that has not been converted to canonical form. In particular, text line breaks must be converted into CRLF sequences prior to base-64 encoding. The important thing to note is that this may be done directly by the encoder rather than in a prior cannibalization step in some implementations.

NOTE: There is no need to worry about quoting apparent encapsulation boundaries within base-64-encoded parts of multipart because no hyphen characters are used in the base-64 encoding.

Annex D - JPEG file interchange format
(normative)

Version 1.02
September 1, 1992

1 408 944-6300
Fax: +1 408 944-6314
E-mail: eric@c3.pla.ca.us

Why a file interchange format

JPEG File Interchange Format (JFIF) is a minimal file format, which enables JPEG bitstreams to be exchanged between a wide variety of platforms and applications. This minimal format does not include any of the advanced features found in the TIFF JPEG specification or any application specific file format. The only purpose of this simplified format is to allow the exchange of JPEG compressed images.

JPEG file interchange format

- Uses JPEG compression
- Uses JPBG interchange format compressed image representation
- PC or Mac or UNIX workstation compatible
- Standard color space: one or three components. For three components YCbCr (CCIR 601-256 levels)
- APP0 marker used to specify Units, X pixel density, Y pixel density, thumbnail
- APP0 marker also used to specify JFIF extensions
- APP0 mater also used to specify application-specific information

JPEG compression

Although any JPEG process is supported by the syntax of the JFIF it is strongly recommended that the JPEG baseline process be used for the purposes of file interchange. This ensures maximum compatibility with all applications supporting JPEG. JFIF conforms to the JPEG Draft International Standard (ISO DIS 10918-1).

The JFIF is entirely compatible with the standard JPEG interchange format; the only additional requirement is the mandatory presence of the APP0 marker right after the SOI marker. Note that the JPEG interchange format requires (as does JFIF) all table specifications used in the encoding process be coded in the bitstream prior to their use.

Compatible across platforms

The JFIF is compatible across platforms: for example, it can use any resource forks supported by the Macintosh and by PCs or workstations, but not just one platform.

Standard color space

The color space to be used is YCbCr as defined by CCIR 601(256 levels). The RGB components calculated by linear conversion from YCbCr shall not be gamma corrected (gamma = 1.0). If only one component is used, that component shall be Y.

APP0 marker is used to identify JPEG FIF

- The APP0 marker is used to identify a JPEG FIF file.
- The JPEG FIF APP0 marker is mandatory right after the SOI marker.
- The JFIF APP0 marker is identified by a zero terminated string: *"JFIF"*.
- The APP0 can be used for any other purpose by the application provided it can be distinguished from the JFIF APP0.
- The JFIF APP0 marker provides information which is missing from the JPEG stream: version number, X and Y pixel density (dots per inch or dots per cm), pixel aspect ratio (derived from X and Y pixel density), thumbnail.

APP0 marker used to specify JFIF extensions

Additional APP0 marker segment(s) can optionally be used to specify JFIF extensions. If used, these segments must immediately follow the JFIF APP0 marker. Decoders should skip any unsupported JFIF extension segments and continue decoding.

The JFIF extension APP0 marker is identified by a zero terminated string: *"JFXX"*. The JFIF extension APP0 marker segment contains a 1-byte code, which identifies the extension. This version, version 1.02, has only one extension defined: an extension for defining thumbnails stored in formats other than 24-bit RGB.

APP0 marker used for application-specific information

Additional APP0 marker segments can be used to hold application-specific information which does not affect the decodability or displayability of the JFIF file. Application-specific APP0 marker segments must appear after the JFIF APP0 and any JFXX APP0 segments. Decoders should skip any unrecognized application-specific APP0 segments.

Application-specific APP0 marker segments are identified by a zero terminated string which identifies the application (not "*JFIF*" or "*JFXX*"). This string should be an organization name or company trademark. Generic strings such as dog, cat, tree, etc. should not be used.

Conversion to and from RGB

Y, Cb, and Cr are converted from R, G, and B as defined in CCIR Recommendation 601 but are normalized so as to occupy the full 256 levels of an 8-bit binary encoding. More precisely:

$Y = 256 * E'_y$
$Cb = 256 * [\ E'_{Cb}] + 128$
$Cr = 256 * [\ E'_{Cr}] + 128$

where the E'_y E'_{Cb} and E'_{Cr} are defined as in CCIR 601. Since values of E'_y have a range of 0 to 1.0 and those for E'_{Cb} and E'_{Cr} have a range of -0.5 to +0.5, Y, Cb, and Cr must be clamped to 255 when they are maximum value.

RGB to YCbCr conversion

YCbCr (256 levels) can be computed directly from 8-bit RGB as follows:

$Y = 0.299\ R + 0.587 G + 0.114 B$
$Cb = -0.1687\ R - 0.3313\ G + 0.5\ B + 128$
$Cr = 0.5 R - 0.4177\ G - 0.0813\ B + 128$

NOTE - Not all image file formats store image samples in the order $R_0, G_0, B_0, ... R_n, G_n, B_n$. Be sure to verify the sample order before converting an RGB file to JFIF

YCbCr to RGB conversion

RGB can be computed directly from YCbCr (256 levels) as follows:

$R = Y + 1.402\ (Cr - 128)$
$G = Y - 0.34414\ (Cb - 128) - 0.71414\ (Cr - 128)$
$B = Y + 1.772\ (Cb - 128)$

Image orientation

In JFIF files, the image orientation is always top-down. This means that the first image samples encoded in a JFIF file are located in the upper left hand corner of the image and encoding proceeds from left to right and top to bottom. Top-down orientation is used for both the full resolution image and the thumbnail image.

The process of converting an image file having bottom-up orientation to JFIF must include inverting the order of all image lines before JPEG encoding.

Spatial relationship of components

Specification of the spatial positioning of pixel samples within components relative to the samples of other components is necessary for proper image post processing and accurate image presentation. In JFIF files, the position of the pixels in subsampled components are defined with respect to the highest resolution component. Since components must be sampled orthogonally (along rows and columns), the spatial position of the samples in a given subsampled component may be determined by specifying the horizontal and vertical offsets of the first sample, i.e. the sample in the upper left corner, with respect to the highest resolution component.

The horizontal and vertical offsets of the first sample in a subsampled component, $Xoffset_i[0,0]$ and $Yoffset_i[0,0]$, are defined to be:

$$Xoffset_i[0,0] = ((Nsamples_{ref} / Nsamples_j) / 2) - 0.5$$

$$Yoffset_i[0,0] = ((Nlines_{ref} / Nlines_i) / 2) - 0.5$$

where

$Nsamples_{ref}$ is the number of samples per line in the largest component;

$Nsamples_i$ is the number of samples per line in the ith component;

$Nlines_{ref}$ is the number of lines in the largest component;

$Nlines_i$ is the number of lines in the ith component.

Proper subsampling of components incorporates an anti-aliasing filter which reduces the spectral bandwidth of the full resolution components. Sub-sampling can easily be accomplished using a symmetrical digital filter with an even number of taps (coefficients). A commonly used filter for 2:1 subsampling utilizes two taps (1/2, 1/2).

As an example, consider a 3 component image which is comprised of components having the following dimensions:

Component 1: 256 samples, 288 lines
Component 2: 128 samples, 144 lines
Component 3: 64 samples, 96 lines

In a JFIF file, centers of the samples are positioned as illustrated below:

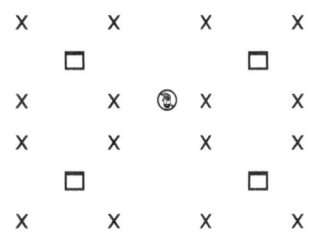

where

X   Component 1
□   Component 2
⊛   Component 3

NOTE - This definition is compatible with industry standards such as Postscript Level 2 and Quick-Time. This definition is <u>not</u> compatible with the conventions used by CCIR Recommendation 601-I and other digital video formats. For these formats, pre-processing of the chrominance components is necessary prior to compression in order to ensure accurate reconstruction of the compressed image.

JPEG file interchange format specification

The syntax of a JFIF file conforms to the syntax for interchange format defined in Annex B of ISO DIS 10918-1. In addition, a JFIF file uses APP0 marker segments and constrains certain parameters in the frame header as defined below.

X'FF', SOI

X'FF', APP0, length, identifier, version, units, Xdensity, Ydensity, Xthumbnail, Ythumbnail, $(RGB)_n$

Length	(2 bytes)	Total APP0 field byte count, including the byte count value (2 bytes), but excluding the APP0 marker itself
identifier	(5 bytes)	= X'4A', X'46', X'49', X'46', X'00' This zero terminated string ("JFIF") Uniquely identifies this APP0 marker. This string shall have zero parity (bit 7=0).
version	(2 bytes)	= X'0102' The most significant byte is used for major revisions, the least significant byte for minor revisions. Version 1.02 is the current released revision.

ANSI/NIST-ITL 1-2000

units	(1 byte)	Units for the X and Y densities units = 0: no units, X and Y specify the pixel units = 1: X and Y are dots per inch units = 2: X and Y are dots per cm
Xdensity	(2 bytes)	Horizontal pixel density
Ydensity	(2 bytes)	Vertical pixel density
Xthumbnail	(1 byte)	Thumbnail horizontal pixel count
Ythumbnail	(1 byte)	Thumbnail vertical pixel count
$(RGB)_n$	(3n bytes)	Packed (24-bit) RGB values for the thumbnail pixels, n = Xthumbnail * Ythumbnail

[Optional JFIF extension APP0 marker segment(s) - see below]
-
-
-

X 'FF', SOFn, length,. frame parameters

Number of components	Nf	= 1 or 3
1st component	C1	= 1 = Y component
2nd component	C2	= 2 = Cb component
3rd component	C3	= 3 = Cr component

-
-
-

X 'FF', EOI

JFIF Extension: APP0 marker segment

Immediately following the JFIF APP0 marker segment may be a JFIF extension APP0 marker. This JFIF extension APP0 marker segment may only be present for JFIF versions 1.02 and above. The syntax of the JFIF extension APP0 marker segment is:

X 'FF', APP0,. Length, identifier, extension code, extension data

length	(2 bytes)	Total APP0 field byte count, including the byte count value (2 bytes), but excluding the APP0 marker itself
identifier	(5 bytes)	= X '4A', X '46'[1], X '58', X '58', X '00' This zero terminated string ("JFXX") uniquely identifies this APP0 marker. This string shall have zero parity (bit 7 = 0).
extension_code	(1 byte)	= Code which identifies the extension. In this version, the following extensions are defined: = X '10' Thumbnail coded using JPEG = X '11' Thumbnail stored using 1 byte/pixel = X '13' Thumbnail stored using 3 bytes/pixel

extension_data   (variable)   = The specification of the remainder of the
                                   JFIF extension APP0 marker segment
                                   varies with the extension. See below for a
                                   specification of extension_data for each extension.

JFIF Extension: Thumbnail coded using JPEG

This extension supports thumbnails compressed using JPEG. The compressed thumbnail immediately follows the extension-code (X '10') in the extension_data field and the length of the compressed data must be included in the JFIF extension APP0 marker length field.

The syntax of the extension_data field conforms to the syntax for interchange format defined in Annex B of ISO DIS 10917-1. However, no "JFIF" or "JFXX" marker segments shall be present. As in the full resolution image of the JFIF file, the syntax of extension_data constrains parameters in the frame header as defined below:

X 'FF', SOI
-
-

X'FF', SOF_n, length, frame parameters

  Number of components Nf = 1 or 3
  1st component C_1 = 1 = Y component
  2nd component C_2 = 2 = Cb component
  3rd component C_3 = 3 = Cr component

-
-

X 'FF', EOI

JFIF Extension: Thumbnail stored using one byte per pixel

This extension supports thumbnails stored using one byte per pixel and a color palette in the extension_data field. The syntax of extension_data is:

Xthumbnail	(l byte)	Thumbnail horizontal pixel count
Ythumbnail	(1 byte)	Thumbnail vertical pixel count
Palette	(768 bytes)	24-bit RGB pixel values for the color palette. The RGB values define the colors represented by each value of an 8-bit binary encoding (0 - 255).
$(pixel)_n$	(n bytes)	8-bit values for the thumbnail pixels n = Xthumbnail * Ythumbnail

JFIF Extension: Thumbnail stored using three bytes per pixel

This extension supports thumbnails stored using three bytes per pixel in the extension_data field. The syntax of extension_data is:

Xthumbnail	(1 byte)	Thumbnail horizontal pixel count
Ythumbnail	(1 byte)	Thumbnail vertical pixel count
$(RGB)_n$	(3n bytes)	Packed (24-bit) RGB values for the thumbnail pixels, n = Xthumbnail * Ythumbnail

ANSI/NIST-ITL 1-2000

Useful tips

- You can identify a JFIF file by looking for the following sequence: *X'FF', SOI, X'FF', APP0*, <2 bytes to be skipped>, *"JFIF", X'00'*.
- If you use APP0 elsewhere, be sure not to have the strings "*JFIF*" or "*JFXX*" right after the APP0 marker.
- If you do not want to include a thumbnail, just program Xthumbnail = Ythumbnail = 0.
- Be sure to check the version number in the special APP0 field. In general, if the major version number of the JFIF file matches that supported by the decoder, the file will be decodable.
- If you only want to specify a pixel aspect ratio, put 0 for the units field in the special APP0 field. Xdensity and Ydensity can then be programmed for the desired aspect ratio. Xdensity = 1, Ydensity = 1 will program a 1:1 aspect ratio. Xdensity and Ydensity should always be non-zero.

ANSI/NIST-ITL 1-2000

Annex E - Scars, marks, tattoos, and other characteristics
(informative)

This annex contains excerpts from Section 13 Part 4 of the Eighth Edition (July 14, 1999) of the NCIC Code Manual for describing Scars, Marks, Tattoos, and other characteristics. The following list is intended to standardize entry of data in the SMT Field. Only the following codes are to be entered in this field. Care must be taken to enter spaces exactly as shown.

Other Physical Characteristics

Item/Location	Code
Bald/Balding	BALD
Cleft chin	CLEFT CHIN
Dimple, Chin	DIMP CHIN
Dimples, left cheek (face)	DIMP L CHK
Dimples, right cheek (face)	DIMP R CHK
Freckles	FRECKLES
Hair implants	HAIR IMPL
Pierced abdomen	PRCD ABDMN
Pierced back	PRCD BACK
Pierced ear, one, nonspecific	PRCD EAR
Pierced ears	PRCD EARS
Pierced left ear	PRCD L EAR
Pierced right ear	PRCD R EAR
Pierced eyebrow, nonspecific	PRCD EYE
Pierced left eyebrow	PRCD L EYE
Pierced right eyebrow	PRCD R EYE
Pierced genitalia	PRCD GNTLS
Pierced lip, nonspecific	PRCD LIP
Pierced lip, upper	PRCD ULIP
Pierced lip, lower	PRCD LLIP
Pierced nipple, nonspecific	PRCD NIPPL
Pierced nipple, left	PRCD L NIP
Pierced nipple, right	PRCD R NIP
Pierced nose	PRCD NOSE
Pierced tongue	PRCD TONGU
Stutters	STUTTERS
Transsexual	TRANSSXL

(Miscellaneous Field should indicate what the individual was at birth and what they are at the time the record is entered into NCIC. [Example: Born male - had surgery and is now female.])

Transvestite	TRANSVST

Scars (SC)

Item/Location	Code
Abdomen	SC ABDOM
Ankle, nonspecific	SC ANKL
Ankle, left	SC L ANKL
Ankle, right	SC R ANKL
Arm, nonspecific	SC ARM
Arm, left	SC L ARM
Arm, right	SC R ARM

(Use the MIS Field to further describe location)

Arm, left upper	SC UL ARM
Arm, right upper	SC UR ARM
Back	SC BACK
Breast, nonspecific	SC BREAST
Breast, left	SC L BRST
Breast, right	SC R BRST
Buttock, nonspecific	SC BUTTK
Buttock, left	SC L BUTTK
Buttock, right	SC R BUTTK

Item/Location	Code	Item/Location	Code
Calf, nonspecific	SC CALF	Knee, left	SC L KNEE
Calf, left	SC L CALF	Knee, right	SC R KNEE
Calf, right	SC R CALF	Leg, nonspecific	SC LEG
		Leg, left, nonspecific	SC L LEG
Cheek (face), nonspecific	SC CHK	Leg, right, nonspecific	SC R LEG
Cheek (face), left	SC L CHK	(Use the MIS Field to further describe location)	
Cheek (face), right	SC R CHK		
		Lip, nonspecific	SC LIP
Chest	SC CHEST	Lip, lower	SC LOW LIP
		Lip, upper	SC UP LIP
Chin	SC CHIN		
		Neck	SC NECK
Ear, nonspecific	SC EAR		
Ear, left	SC L EAR	Nose	SC NOSE
Ear, right	SC R EAR		
		Penis	SC PENIS
Elbow, nonspecific	SC ELBOW		
Elbow, left	SC L ELB	Pockmarks	POCKMARKS
Elbow, right	SC R ELB		
		Shoulder, nonspecific	SC SHLD
Eyebrow, nonspecific	SC EYE	Shoulder, left	SC L SHLD
Eyebrow, left/left eye area	SC L EYE	Shoulder, right	SC R SHLD
Eyebrow, right/right eye area	SC R EYE		
		Thigh, nonspecific	SC THGH
Face, nonspecific	SC FACE	Thigh, left	SC L THGH
(Use the MIS Field to further describe location)		Thigh, right	SC R THGH
		Wrist, nonspecific	SC WRIST
Finger, nonspecific	SC FGR	Wrist, left	SC L WRIST
Finger(s), left hand	SC L FGR	Wrist, right	SC R WRIST
Finger(s), right hand	SC R FGR		
Foot, nonspecific	SC FOOT		
Foot, left	SC L FT		
Foot, right	SC R FT	**Needle ("Track") Marks (NM)**	
Forearm, nonspecific	SC F ARM	**Item/Location**	**Code**
Forearm, left	SC LF ARM	Arm, left	NM L ARM
Forearm, right	SC RF ARM	Arm, right	NM R ARM
Forehead	SC FHD	Buttock, left	NM L BUTTK
		Buttock, right	NM R BUTTK
Groin area	SC GROIN		
		Finger(s), left hand	NM L FGR
Hand, nonspecific	SC HAND	Finger(s), right hand	NM R FGR
Hand, left	SC L HND		
Hand, right	SC R HND	Foot, left	NM L FOOT
Head, nonspecific	SC HEAD	Foot, right	NM R FOOT
(Use the MIS Field to further describe location)			
		Hand, left	NM L HND
Hip, nonspecific	SC HIP	Hand, right	NM R HND
Hip, left	SC L HIP		
Hip, right	SC R HIP	Leg, left	NM L LEG
		Leg, right	NM R LEG
Knee, nonspecific	SC KNEE		

Item/Location	Code
Thigh, left	NM L THIGH
Thigh, right	NM R THIGH
Wrist, left	NM L WRIST
Wrist, right	NM R WRIST

Tattoos (TAT)

Item/Location	Code
Abdomen	TAT ABDOM
Ankle, nonspecific	TAT ANKL
Ankle, left	TAT L ANKL
Ankle, right	TAT R ANKL
Arm, nonspecific	TAT ARM
Arm, left	TAT L ARM
Arm, right	TAT R ARM

(Use the MIS Field to further describe location)

Arm, left upper	TAT UL ARM
Arm, right upper	TAT UR ARM
Back	TAT BACK
Breast, nonspecific	TAT BREAST
Breast, left	TAT L BRST
Breast, right	TAT R BRST
Buttocks, nonspecific	TAT BUTTK
Buttock, left	TAT L BUTK
Buttock, right	TAT R BUTK
Calf, nonspecific	TAT CALF
Calf, left	TAT L CALF
Calf, right	TAT R CALF
Cheek (face), nonspecific	TAT CHEEK
Cheek (face), left	TAT L CHK
Cheek (face), right	TAT R CHK
Chest	TAT CHEST
Chin	TAT CHIN
Ear, nonspecific	TAT EAR
Ear, left	TAT L EAR
Ear, right	TAT R EAR
Elbow, nonspecific	TAT ELBOW
Elbow, left	TAT LELBOW
Elbow, right	TAT RELBOW
Face, nonspecific	TAT FACE

(Use the MIS Field to further describe location)

Finger, nonspecific	TAT FNGR
Finger(s), left hand	TAT L FGR
Finger(s), right hand	TAT R FGR
Foot, nonspecific	TAT FOOT
Foot, left	TAT L FOOT
Foot, right	TAT R FOOT
Forearm, nonspecific	TAT FARM
Forearm, left	TAT LF ARM
Forearm, right	TAT RF ARM
Forehead	TAT FHD
Full Body	TAT FLBODY

(Use only when the entire body - arms, legs, chest, and back are covered with tattoos)

Groin Area	TAT GROIN
Hand, nonspecific	TAT HAND
Hand, left	TAT L HND
Hand, right	TAT R HND
Head, nonspecific	TAT HEAD

(Use the MIS Field to further describe location)

Hip, nonspecific	TAT HIP
Hip, left	TAT L HIP
Hip, right	TAT R HIP
Knee, nonspecific	TAT KNEE
Knee, left	TAT L KNEE
Knee, right	TAT R KNEE
Leg, nonspecific	TAT LEG
Leg, left	TAT L LEG
Leg, right	TAT R LEG

(Use the MIS Field to further describe location)

Lip, nonspecific	TAT LIP
Lip, lower	TAT LW LIP
Lip, upper	TAT UP LIP
Neck	TAT NECK
Nose	TAT NOSE
Penis	TAT PENIS

ANSI/NIST-ITL 1-2000

Item/Location	Code
Shoulder, nonspecific	TAT SHLD
Shoulder, left	TAT L SHLD
Shoulder, right	TAT R SHLD
Thigh, nonspecific	TAT THGH
Thigh, left	TAT L THGH
Thigh, right	TAT R THGH
Wrist, nonspecific	TAT WRS
Wrist, left	TAT L WRS
Wrist, right	TAT R WRS

Removed Tattoos (RTAT)

Item/Location	Code
Abdomen	RTAT ABDM
Ankle, nonspecific	RTAT ANKL
Ankle, left	RTAT LANKL
Ankle, right	RTAT RANKL
Arm, nonspecific	RTAT ARM
Arm, left	RTAT L ARM
Arm, right	RTAT R ARM

(Use the MIS Field to further describe location)

Arm, left upper	RTAT ULARM
Arm, right upper	RTAT URARM
Back	RTAT BACK
Breast, nonspecific	RTAT BRST
Breast, left	RTAT LBRST
Breast, right	RTAT RBRST
Buttocks, nonspecific	RTAT BUTTK
Buttock, left	RTAT LBUTK
Buttock, right	RTAT RBUTK
Calf, nonspecific	RTAT CALF
Calf, left	RTAT LCALF
Calf, right	RTAT RCALF
Cheek (face), nonspecific	RTAT CHEEK
Cheek (face), left	RTAT L CHK
Cheek (face), right	RTAT R CHK
Chest	RTAT CHEST
Chin	RTAT CHIN
Ear, nonspecific	RTAT EAR

Item/Location	Code
Ear, left	RTAT L EAR
Ear, right	RTAT R EAR
Elbow, nonspecific	RTAT ELBOW
Elbow, left	RTAT L ELB
Elbow, right	RTAT R ELB
Face, nonspecific	RTAT FACE

(Use the MIS Field to further describe location)

Finger, nonspecific	RTAT FNGR
Finger(s), left hand	RTAT L FGR
Finger(s), right hand	RTAT R FGR
Foot, nonspecific	RTAT FOOT
Foot, left	RTAT LFOOT
Foot, right	RTAT RFOOT
Forearm, nonspecific	RTAT FARM
Forearm, left	RTAT LFARM
Forearm, right	RTAT RFARM
Forehead	RTAT FHD
Full body	RTAT FLBOD
Groin area	RTAT GROIN
Hand, nonspecific	RTAT HAND
Hand, left	RTAT L HND
Hand, right	RTAT R HND
Head, nonspecific	RTAT HEAD

(Use the MIS Field to further describe location)

Hip, nonspecific	RTAT HIP
Hip, left	RTAT L HIP
Hip, right	RTAT R HIP
Knee, nonspecific	RTAT KNEE
Knee, left	RTAT LKNEE
Knee, right	RTAT RKNEE
Leg, nonspecific	RTAT LEG
Leg, left	RTAT L LEG
Leg, right	RTAT R LEG

(Use the MIS Field to further describe location)

Lip, nonspecific	RTAT LIP
Lip, lower	RTAT LWLIP
Lip, upper	RTAT UPLIP
Neck	RTAT NECK
Nose	RTAT NOSE

Penis	RTAT PENIS
Shoulder, nonspecific	RTAT SHLD
Shoulder, left	RTAT LSHLD
Shoulder, right	RTAT RSHLD
Thigh, nonspecific	RTAT THGH
Thigh, left	RTAT LTHGH
Thigh, right	RTAT RTHGH
Wrist, nonspecific	RTAT WRS
Wrist, left	RTAT LWRS
Wrist, right	RTAT RWRS

ANSI/NIST-ITL 1-2000

Annex F - Example transaction
(informative)

This annex contains an example of the use of the standard for the interchange of specific types of biometric image data between different systems or organizations. The TOT code is fictional and used to represent a transaction for submitting fingerprint, mugshot, and palmprint image data to a remote system. In addition to the Type-1 record, this transaction includes eighteen other records. It contains a Type-2 user-defined descriptive text record, fourteen Type-14 variable resolution tenprint image records, a Type-10 mugshot record, and two Type-15 variable resolution records containing a full and writer's palmprint image. A scanning resolution of 19.69 ppmm (500 ppi) was used for the fingerprint and palmprint images. The WSQ Version 2.0 compression algorithm was used to compress these images to approximately 15:1. For the mugshot image, a JPEG Baseline algorithm was used to compress the image to approximately 20:1.

Type-1 Transaction Record

LENGTH (LEN)	1.001: 305 $_S^G$
VERSION (VER)	1.002:0300 $_S^G$
CONTENT (CNT)	1.003:1 $_S^U$ 18 $_S^R$ 2 $_S^U$ 00 $_S^R$ 14 $_S^U$ 01 $_S^R$ 14 $_S^U$ 02 $_S^R$ 14 $_S^U$ 03 $_S^R$ 14 $_S^U$ 04 $_S^R$
	14 $_S^U$ 05 $_S^R$ 14 $_S^U$ 06 $_S^R$ 14 $_S^U$ 07 $_S^R$ 14 $_S^U$ 08 $_S^R$ 14 $_S^U$ 09 $_S^R$ 14 $_S^U$ 10 $_S^R$ 14 $_S^U$
	11 $_S^R$ 14 $_S^U$ 12 $_S^R$ 14 $_S^U$ 13 $_S^R$ 14 $_S^U$ 14 $_S^R$ 10 $_S^U$ 15 $_S^R$ 15 $_S^U$ 16 $_S^R$ 15 $_S^U$ 17 $_S^G$
TYPE OF TRANSACTION (TOT)	1.004:XXX $_S^G$
DATE (DAT)	1.005:19991120 $_S^G$
PRIORITY (PRY)	1.006:1 $_S^G$
DESTINATION AGENCY IDENTIFIER (DAI)	1.007:DCFBIWA6Z $_S^G$
ORIGINATING AGENCY IDENTIFIER (ORI)	1.008:NY0303000SLAS01000 $_S^G$
TRANSACTION CONTROL NUMBER (TCN)	1.009:1234567890 $_S^G$
TRANSACTION CONTROL REFERENCE (TCR)	1.010:2345678901 $_S^G$
NATIVE SCANNING RESOLUTION (NSR)	1.011:19.69 $_S^G$

ANSI/NIST-ITL 1-2000

TRANSMITTING RESOLUTION (NTR)	1.012:19.69 $_S^G$
DOMAIN NAME (DOM)	1.013:NORAM $_S^U$ $_S^G$
GREENWICH MEAN TIME (GMT)	1.014:19991120235745Z $_S^F$

Type-2 User-Defined Descriptive Text Record

LENGTH (LEN)	2.001:813 $_S^G$
IMAGE DESIGNATION CHARACTER (IDC)	2.002:00 $_S^G$
USER-DEFINED INFORMATION	(793 ASCII TEXT CHARACTERS) $_S^F$

Type-14 Variable-Resolution Tenprint Records

1ST TYPE-14 RECORD (RIGHT THUMB)—

LOGICAL RECORD LENGTH (LEN)	14.001:40164 $_S^G$
IMAGE DESIGNATION CHARACTER (IDC)	14.002:01 $_S^G$
IMPRESSION TYPE (IMP)	14.003:1 $_S^G$
SOURCE AGENCY (SRC)	14.004:NY0303000SLA0100 $_S^G$
CAPTURE DATE (TCD)	14.005:19991120 $_S^G$
HORIZONTAL LINE LENGTH (HLL)	14.006:800 $_S^G$
VERTICAL LINE LENGTH (VLL)	14.007:750 $_S^G$
SCALE UNITS (SLC)	14.008:2 $_S^G$
HORIZONTAL PIXEL SCALE (HPS)	14.009:197 $_S^G$
VERTICAL PIXEL SCALE (VPS)	14.010:197 $_S^G$
COMPRESSION ALGORITHM (CGA)	14.011:WSQ20 $_S^G$

ANSI/NIST-ITL 1-2000

BITS PER PIXEL (BPX)	14.012:8 ^G_S
FINGER POSITION (FGP)	14.013:1 ^G_S
IMAGE DATA (DAT)	14.999: <40KB OF DATA COMPRESSED @ 15:1> ^F_S

<u>2ND TYPE-14 RECORD (RIGHT INDEX)—</u>

LOGICAL RECORD LENGTH (LEN)	14.001:40164 ^G_S
IMAGE DESIGNATION CHARACTER (IDC)	14.002:02 ^G_S
IMPRESSION TYPE (IMP)	14.003:1 ^G_S

-
-
-

<u>14TH TYPE-14 RECORD (PLAIN LEFT FOUR)—</u>

LOGICAL RECORD LENGTH (LEN)	14.001:107168 ^G_S
IMAGE DESIGNATION CHARACTER (IDC)	14.002:14 ^G_S
IMPRESSION TYPE (IMP)	14.003:0 ^G_S
SOURCE AGENCY (SRC)	14.004:NY0303000SLA0100 ^G_S
CAPTURE DATE (TCD)	14.005:19991120 ^G_S
HORIZONTAL LINE LENGTH (HLL)	14.006:1600 ^G_S
VERTICAL LINE LENGTH (VLL)	14.007:1000 ^G_S
SCALE UNITS (SLC)	14.008:2 ^G_S
HORIZONTAL PIXEL SCALE (HPS)	14.009:197 ^G_S
VERTICAL PIXEL SCALE (VPS)	14.010:197 ^G_S
COMPRESSION ALGORITHM (CGA)	14.011:WSQ20 ^G_S
BITS PER PIXEL (BPX)	14.012:8 ^G_S

ANSI/NIST-ITL 1-2000

FINGER POSITION (FGP)	14.013:14 ^G_S
IMAGE DATA (DAT)	14.999: <107KB OF DATA COMPRESSED @ 15:1> ^F_S

Type-10 Facial / Mugshot Record

LENGTH (LEN)	10.001:14601 ^G_S
IMAGE DESIGNATION CHARACTER (IDC)	10.002:15 ^G_S
IMAGE TYPE (IMT)	10.003:FACE ^G_S
SOURCE AGENCY (SRC)	10.004:NY0303000SLAS01000 ^G_S
*PHOTO DATE (PHD)	10.005:19991120 ^G_S
HORIZONTAL LINE LENGTH (HLL)	10.006:480 ^G_S
VERTICAL LINE LENGTH (VLL)	10.007:600 ^G_S
SCALE UNITS (SLC)	10.008:0 ^G_S
HORIZONTAL PIXEL SCALE (HPS)	10.009:1 ^G_S
VERTICAL PIXEL SCALE (VPS)	10.010:1 ^G_S
COMPRESSION ALGORITHM (CGA)	10.011:JPEGB ^G_S
COLORSPACE (CSP)	10.012:YCC ^G_S
SUBJECT POSE (POS)	10.020:A ^G_S
POSE OFFSET ANGLE (POA)	10.021:-45 ^G_S
PHOTO DESCRIPTION (PXS)	10.022:HAT ^R_S GLASSES ^G_S
IMAGE DATA (DAT)	10.999:
SOI & APP0 Marker Segment	X'FFD8', X'FFE0' X'0010' 4A46494600', X'0102', X'00', X'0001', X'0001', X'00', X'00',

ANSI/NIST-ITL 1-2000

Compressed Image Data	<14382 BYTES OF FACIAL DATA COMPRESSED @ 20:1>
End Of Image Marker Code	X'FFD9' ^F_S

Type-15 Variable-Resolution Palmprint Records

1ST TYPE-15 RECORD (FULL RIGHT PALM)—

LENGTH (LEN)	15.001:733170 ^G_S
IMAGE DESIGNATION CHARACTER (IDC)	15.002:16 ^G_S
IMPRESSION TYPE (IMP)	15.003:0 ^G_S
SOURCE AGENCY (SRC)	15.004:NY0303000SLAS01000 ^G_S
PALMPRINT CAPTURE DATE (PCD)	15.005:19991120 ^G_S
HORIZONTAL LINE LENGTH (HLL)	15.006:2750 ^G_S
VERTICAL LINE LENGTH (VLL)	15.007:4000 ^G_S
SCALE UNITS (SLC)	15.008:2 ^G_S
HORIZONTAL PIXEL SCALE (HPS)	15.009:197 ^G_S
VERTICAL PIXEL SCALE (VPS)	15.010:197 ^G_S
COMPRESSION ALGORITHM (CGA)	15.011:WSQ20 ^G_S
BITS PER PIXEL (BPX)	15.012:8 ^G_S
PALMPRINT POSITION (PLP)	15.020:21 ^G_S
IMAGE DATA (DAT)	15.999: <733KB OF DATA COMPRESSED @ 15:1> ^F_S

2ND TYPE-15 RECORD (WRITER'S RIGHT PALM)—

LENGTH (LEN)	15.001:146169 ^G_S
IMAGE DESIGNATION CHARACTER (IDC)	15.002:17 ^G_S

IMPRESSION TYPE (IMP)	15.003:0 ^G_S
SOURCE AGENCY (SRC)	15.004:NY0303000SLAS01000 ^G_S
PALMPRINT CAPTURE DATE (PCD)	15.005:19991120 ^G_S
HORIZONTAL LINE LENGTH (HLL)	15.006:876 ^G_S
VERTICAL LINE LENGTH (VLL)	15.007:2500 ^G_S
SCALE UNITS (SLC)	15.008:2 ^G_S
HORIZONTAL PIXEL SCALE (HPS)	15.009:197 ^G_S
VERTICAL PIXEL SCALE (VPS)	15.010:197 ^G_S
COMPRESSION ALGORITHM (CGA)	15.011:WSQ20 ^G_S
BITS PER PIXEL (BPX)	15.012:8 ^G_S
PALMPRINT POSITION (PLP)	15.020:22 ^G_S
IMAGE DATA (DAT)	15.999: <146KB OF DATA COMPRESSED @ 15:1> ^F_S

www.ingramcontent.com/pod-product-compliance
Lightning Source LLC
Chambersburg PA
CBHW081839170526
45167CB00007B/2842